Surface Processing of Light Alloys Subject to Concentrated Energy Flows

Xizhang Chen · Sergey Konovalov ·
Victor Gromov · Yurii Ivanov

Surface Processing of Light Alloys Subject to Concentrated Energy Flows

Xizhang Chen
Wenzhou University
Wenzhou, China

Sergey Konovalov
Samara National Research University
Samara, Russia

Victor Gromov
Siberian State Industrial University
Novokuznetsk, Russia

Yurii Ivanov
Institute of High Current Electronics,
Siberian Branch
Russian Academy of Science
Tomsk, Russia

ISBN 978-981-33-4227-9   ISBN 978-981-33-4228-6 (eBook)
https://doi.org/10.1007/978-981-33-4228-6

Jointly published with Science Press
The print edition is not for sale in China (Mainland). Customers from China (Mainland) please order the print book from: Science Press.

© Science Press 2021
This work is subject to copyright. All rights are reserved by the Publishers, whether the whole or part of the material is concerned, specifically the rights of translation, reprinting, reuse of illustrations, recitation, broadcasting, reproduction on microfilms or in any other physical way, and transmission or information storage and retrieval, electronic adaptation, computer software, or by similar or dissimilar methodology now known or hereafter developed.
The use of general descriptive names, registered names, trademarks, service marks, etc. in this publication does not imply, even in the absence of a specific statement, that such names are exempt from the relevant protective laws and regulations and therefore free for general use.
The publishers, the authors, and the editors are safe to assume that the advice and information in this book are believed to be true and accurate at the date of publication. Neither the publishers nor the authors or the editors give a warranty, express or implied, with respect to the material contained herein or for any errors or omissions that may have been made. The publishers remain neutral with regard to jurisdictional claims in published maps and institutional affiliations.

This Springer imprint is published by the registered company Springer Nature Singapore Pte Ltd.
The registered company address is: 152 Beach Road, #21-01/04 Gateway East, Singapore 189721, Singapore

Foreword

I am glad to write the foreword for the 1st edition of the book, *Surface Processing of Light Alloys Subject to Concentrated Energy Flows*. The authors, Xizhang Chen, Sergey Konovalov, Victor Gromov, and Yurii Ivanov have done an excellent compilation of their international collaborative arid cooperative work on enhancing surface properties of light metal alloys via external energy sources. Their works are interdisciplinary in nature that requires in-depth understanding of solid-state physics, materials science, instrumentation, mechanical engineering, and metallurgy. The ability to engineer excellent surfaces depends on synergistic knowledge of these fields. The authors clearly establish the fact that by modifying surfaces of light metals/alloys, their microstructure and eventually their mechanical properties can be controlled, without resorting to transformation of the bulk structure. By utilizing high energy methods such as pulsed electron-beam irradiation and electrical explosion alloying, they have demonstrated change in microstructural evolution that has enhanced the mechanical properties of light-weight materials such as titanium and aluminum–silicon alloys. In particular, prominence has been given to fatigue behavior considering the importance of these materials in applications related to space/aerospace and automotive sectors. The most outstanding feature of this book is the detailed presentation on the significant enhancement of material properties by the combination of electrical explosion alloying and high energy electron-beam processing, which can bring forth tremendous change in microstructure and mechanical properties when compared to those that can be attained by using either one of the two processes. The chapters in the book have been logically arranged. Each chapter presents detailed scientific discussion supported by technical data and microscopy evidences.

The book begins with an initial chapter on the modification of microstructure and properties of metals by concentrated energy flows. An excellent background on fatigue failure in metals and alloys sets the platform upon which the authors build their scientific presentations in the rest of the chapters. Here, a short discussion on effect of electron-beam processing on the fatigue strength of various steels is also presented. Next, details on methods of fatigue tests, electron-beam processing, and electrical explosion alloying are presented. What is commendable here are the details given on the critical parameters of electron-beam unit and the design of electrical explosion apparatus. This will be very valuable information for the scientific community. From here, the authors proceed systematically to discuss the effect of (i) high energy beam processing and (ii) electrical explosion alloying on the microstructural evolution and fatigue properties of aluminum–silicon (silumin) and titanium alloys. It is appreciable that the electrical explosion alloying has been done on titanium metal/alloys with several alloying materials that include titanium diboride, boron carbide, and silicon carbide. Microhardness and wear resistance of the surface-modified titanium materials have also been discussed in terms of role of powder portion weight and surface energy density. The influence of these two parameters on the depth-wise microhardness distribution has been addressed. Lastly, the authors present the combined effects of both electrical explosion alloying and high energy beam processing on the microstructural evolution and mechanical properties of the light-weight materials. The book includes studies pertaining to the mechanisms related to the evolution and formation of structure-phase states, phase composition and dislocation sub-structures in the alloys, which provide important insights into the physical nature of the transformations, in order to achieve high strength and plasticity in the materials. The knowledge and experience of the authors in this field is evident in the entire book.

In this information age, availability of scientific information is very crucial. By sharing their findings from their research works; the authors have made a great contribution to the scientific community. The transdisciplinary aspect of the book written in simple language and in a practical way will appeal to scholars as well as industry personnel worldwide. This book is a must for scientists, engineers, and researchers who are in the quest of novel methods to control surface properties of engineering materials for a variety of applications. I am confident that this book will ignite right scientific temper and will bring several breakthroughs in the domain of materials and surface engineering. The book will excite, motivate, and educate the current and next-generation researchers in advancing the field.

Fedor V. Grechnikov
Academician of the Russian Academy of Sciences
Samara, Russia

Preface

Promotion of advanced engineering technologies demands new structural materials with high-grade processing and performance characteristics. However, it is rarely possible to manufacture and process such materials and products utilizing traditional production technologies. Therefore, a key aspect of current research and engineering policy is the development arid integration of innovative technological processes, i.e. utilizing external energy deposition, with the purpose to influence the plasticity of materials and to increase their resistance to failure.

In this regard, a primary concern of solid-state physics is to explore investigations on the mechanisms related to the evolution and formation of structure-phase states and dislocation sub-structures in steels and alloys, when exposed to external energy sources. In addition, experimental studies on structures and phase composition of external energy deposition products will give important insights into the physical nature of these transformations. Such research will provide valuable data to carry out required modifications of structure and performance characteristics of engineering materials. Moreover, a deep understanding of physical mechanisms and nature of structure-phase transformations at macro-to-nano levels is important to achieve the combination of high strength and plasticity.

Despite several comprehensive studies on the properties of surface-modified light metals and alloys, it is still a prospective research area. The performance of a hard surface depends on various factors, such as (i) the strength and hardness of the hardened zone, (ii) its consistent structure and properties, and (iii) its sufficient resistance to fracture, particularly, to crack initiation. Over the last thirty years, researchers all over the world have been focusing on this issue and making attempts to obtain a solution using external energy sources. From the existing research; it is recognized that these methods are efficient solutions to alter the surface properties of various materials. Such external energy interactions are reported to facilitate transformations in the defect structure and (or) structure-phase state in the subsurface layers.

Several techniques of external energy deposition have been developed so far, e.g. traditional chemical-thermal and thermal-mechanical processing, the usage of laser, plasma, ultrasound, electron, ion beams, etc. Surface modification by means of concentrated energy flows, e.g. laser irradiation, strong electron and ion beams, plasma flows, and jets is known to be cost-effective.

Currently, findings of expert assessment have confirmed that various surface treatment techniques which involve concentrated energy flows have been applied in motorcar, aircraft, and space industries. In addition, these surface treatment techniques can compete successfully with diverse coating processes.

Innovative technologies and efficient instrumentation for surface hardening are moved forward by their potential as well as by well-developed equipment for their implementation.

Studies examining various impulse modifying techniques have demonstrated their usefulness and significance in beneficially modifying the surfaces of several materials. However, each of these methods can solve only a particular problem, and thereby has stimulated the search for new alternative and complementary processes that can enhance each other. One of such easy-to-apply method being developed rigorously in the last decade is the "alloying of surfaces by impulse plasma jets," generated via electrical explosion of conductors.

Electrical explosion alloying involves the melting of the surfaces, while convective processes regulate the spreading of alloying elements over material surfaces. The unique characteristic of electrical explosion alloying is that the multiphase jet of explosion products is the source of alloying elements. The outcomes of the process depend on thermal, force, and chemical factors affecting the surface. A number of studies have investigated the simultaneous and interrelated processes in electrical explosion alloying, which determine the formation of new structure-phase states arid eventually the properties of surfaces. Therefore, it is important to explore the potential of the electrical explosion alloying technique, and also to control its products and develop highly mechanized and special automated equipment. Further, it is required to undertake research to reveal most probable fields of electrical explosion alloying, since, to date, this technique has been adapted only for the melting and formation of a well-developed surface topography. Once the treatment has been completed, the surface will be improved via the supplementary electron-beam processing.

The combination of (i) impulse plasma jets used in electrical explosion alloying and (ii) impulse high-current beams, generated in an experimental unit "SOLO" (Institute of High-Current Electronics, Siberian Branch of the Russian Academy of Sciences). In these two processes, the parameters such as the impulse time, diameter of the surface to be irradiated, the intensity of exposure, and the depth of the processed zone are quite comparable. Further, there is no pressure applied to the surface under electron irradiation, which melts the surface, leveling its topography due to capillary forces as a consequence.

Along with the reasons mentioned above, other factors related to the combination of electrical explosion alloying with subsequent electron-beam processing have been drawing attention of researchers. Apparently, it may change the structure-phase states and properties of surface layers. Several studies investigating this phenomenon have been already carried out. Electron-beam processing allows varying the amount of the supplied energy and extending the zone when the flow of concentrated energy is applied. Besides, it has low coefficient of energy reflection, high concentration of energy in a volume unit of a material, due to which high-degree non-equilibrium states can be created in the material.

Ion beams are also used to modify materials surfaces. When compared with strong ion beams, in electron-beam processing, low-energy (<30 keV) dense electron beams are generated with a considerably higher input–output ratio (< 90%) in a frequency-impulse (to ≈ 10 Hz) mode at lower (by an order of magnitude) accelerating voltages. In addition, no special radiation protection equipment is needed for X-ray radiation as it is screened by the walls of the processing vacuum chamber. Several factors set apart impulse electron beams from impulse flows of low-temperature plasma beams. They include (i) better efficiency of energy and its density in the cross-section of a flow, (ii) sufficient repeatability of impulses, and (iii) their high frequency. So far, certain criteria distinguishing electrical explosion alloying have been highlighted in literature studies. These include: (a) ultrafast heating (up to 10^6 K/s) of the surface to required temperatures, (b) reaching maximal permitted heat gradients (up to 10^7–10^8 K/m), and (c) cooling of the surface layers at speeds of 10^4–10^9 K/s. These are possible because of the heat transfer throughout the material. As a consequence, non-equilibrium sub-micro and nano-crystalline and amorphous structure-phase states can develop in surface layers.

Major problems that require to be solved in terms of operating equipment/facilities are (i) enhancement of their strength and (ii) extending their endurance. The equipment experience mechanical, thermal, electro-magnetic, hydro, and aerodynamic types of reloading leading to cause plastic strain of stressed zones in severe conditions. The most unique and high-duty products, machinery and facilities run in conditions of cyclic deformations, when even low loads might lead to fractures. A long service life and the reliability of equipment are connected with its fatigue resistance for a majority of machine components, which are exposed to dynamic, repeated, and sign-variable loads, and the fatigue failure is considered to be a principal breakdown.

Undoubtedly, low-energy high-current electron beams, being a universal instrument to transform physical and mechanical properties, are able to prolong the fatigue life of metal machine parts. However, researchers have emphasized that this approach toward the control of fatigue criteria needs a diagnosis of fatigue failures as well as data on evaluating structure-phase states, i.e. laws, which determine their interaction with electron beams.

In this book, the structure and properties of light alloys under the influence of external energy source are presented. The book has been divided into 9 chapters, focusing on various aspects related to the main theme of the research.

In Chapter 1, the modification of structure-phase states and properties of metals by external energy flows are discussed. Focus is mainly on the surface hardening and fatigue failure aspects of metals and alloys due to the interaction with the energy source. In Chapter 2, special analysis related to the modified light alloys is presented. Methodologies related to fatigue tests and electron-beam processing are highlighted. Details pertaining to the vacuum impulse electrical explosion equipment are explained. Considering titanium alloys, the equipment required for their processing using low-energy high-current electron beams is mentioned.

In Chapter 3, the structure and properties of Al–Si alloys (silumin) processed by intense pulsed electron beam are discussed. In Chapter 4, the high-cycle fatigue research results of silumin after intense pulsed electron-beam irradiation are discussed in detail, in terms of fractography. The degradation of silumin with regard to its structure and properties after high-cycle fatigue tests is explained in detail in Chapter 5. The evolution of sub-structure defects and phase states are analyzed.

Chapter 6 presents the specifics related to the surface modification of titanium alloy VT6 by electrical explosion alloying, using titanium diboride, boron carbide, and silicon carbide.

Chapter 7 explains the combinatorial approach for the surface modification of titanium alloy VT6 of using both electrical explosion alloying with diboride, boron carbide, and silicon carbide, followed by electron-beam processing.

Chapter 8 deals with the microhardness and wear resistance of modified layers, wherein the depth-wise distribution of microhardness is highlighted. Further, various parameters such as (i) the role of powder portion weight and (ii) the importance of surface energy density, for the depth-wise microhardness distribution are explained. Finally, the wear resistance of modified layers has been presented.

Chapter 9 focuses on the effect of electron-beam processing and the structure and phase composition of titanium VT1-0 after fatigue tests, wherein fractography, evolution of defect sub-structure, structure and phase composition, and developments of gradient structure are discussed. The analysis is focused on pulsed electron-beam processing.

The book gives a comprehensive understanding of the recent trends in structure and properties of light metals and alloys under the influence of external energy sources.

Wenzhou, China Xizhang Chen
Samara, Russia Sergey Konovalov
Novokuznetsk, Russia Victor Gromov
Tomsk, Russia Yurii Ivanov

Contents

1	**Modifying of Structure-Phase States and Properties of Metals by Concentrated Energy Flows**	1
	1.1 Fatigue Failure in Metals and Alloys	1
	1.2 Face Hardening of Metals and Alloys by Concentrated Energy Flows	5
	1.3 The Effect of Electron-Beam Processing on Fatigue Strength of Various Steels	15
	1.4 The Relevance of Face Hardening Methods for the Structure and Properties of Aluminum–Silicon Alloys	21
	1.5 Processing of the Surfaces in Titanium and Titanium-Based Alloys	30
	1.6 The Use of Concentrated Energy Flows for the Face Hardening of Titanium and Its Alloys	33
	1.7 The Modifying of Structure and Properties in a Complex Surface Treatment	36
	References	41
2	**Special Analysis Aspects of Modified Light Alloys**	53
	2.1 Materials of Research	53
	2.2 Methods of Fatigue Tests	53
	2.3 Methods of Electron-Beam Processing	54
	2.4 Vacuum Impulse Electrical Explosion Apparatus EVU 60/10 for the Generation of Impulse Multiphase Plasma Jets	58
	2.5 Equipment for the Processing of Titanium Alloy Surface by Low-Energy High-Current Electron Beam	62
	2.6 Methods of Structural Studies	64
	2.7 Methods of Quantity-Related Proceeding of Research Data	71
	References	72
3	**Structure and Properties of As-Cast Silumin and Processed by Intense Pulsed Electron Beam**	75
	3.1 Structure-Phase Study of As-Cast Silumin	75

	3.2 Structure and Phase Composition of Silumin Irradiated by an Intense Pulsed Electron Beam	80
	References	89
4	**Fractography of Silumin Surface Fractured in High-Cycle Fatigue Tests**	**91**
	4.1 Fractography of a Fatigue Failure Surface in as-Cast Silumin	93
	4.2 Fractography of Fatigue Failure Surface in Silumin Irradiated by an Intense Pulsed Electron Beam	96
	4.2.1 Analysis of a Fracture Surface in Silumin Samples Modified by an Electron Beam with a Minimal Fatigue Life	97
	4.2.2 Analysis of a Fracture Surface in Silumin Samples Modified by an Electron Beam with a Maximal Fatigue Life	101
	References	107
5	**Degradation of Silumin Structure and Properties in High-Cycle Fatigue Tests**	**109**
	5.1 Degradation of Silumin Properties Irradiated by an Electron Beam in High-Cycle Fatigue Tests	109
	5.2 Evolution of Defect Substructure and Phase State of Silumin Irradiated by an Intensive Pulsed Electron Beam During Fatigue Testing	112
	References	121
6	**Modifying of Titanium VT6 Alloy Surface by Electrical Explosion Alloying**	**123**
	6.1 Electrical Explosion Alloying of Titanium VT6 Alloy by Titanium Diboride	123
	6.2 Electrical Explosion Alloying of Titanium Alloy VT6 Surface by Boron Carbide	127
	6.3 Electrical Explosion Alloying of Titanium Alloy VT6 Surface by Silicon Carbide	131
	References	135
7	**Modifying of Titanium Alloy VT6 Surface by Electrical Explosion Alloying and Electron-Beam Processing**	**137**
	7.1 Research into Titanium Alloy VT6 Processed in Electrical Explosion of Diboride and Irradiated by Electron Beam	137
	7.2 Effect of Electron-Beam Processing on Modifying of Titanium Surface Alloyed in an Electrical Explosion by Boron Carbide	145
	7.3 Effect of Electron-Beam Processing on Modifying of Titanium Surface Alloyed in Electrical Explosion by Silicon Carbide	151
	References	158

8 Microhardness and Wear Resistance of Modified Layers ... 161
8.1 Depthwise Distribution of Microhardness in Modified Layers ... 161
8.1.1 Role of Powder Portion Weight for Depthwise Microhardness Distribution in the Zone of Electrical Explosion Alloying ... 161
8.1.2 Importance of Surface Energy Density for Depthwise Microhardness Distribution in Zone Irradiated by Electron Beams ... 163
8.2 Wear Resistance of Modified Layers ... 166
References ... 170

9 Effect of Electron-Beam Processing on Structure and Phase Composition of Titanium VT1-0 Fractured in Fatigue Tests ... 171
9.1 Fracture Surface, Structures, and Phase Composition of Fractured Titanium VT1-0 When Fatigued ... 171
9.1.1 Fractography of the Fatigue Fracture Surface ... 171
9.1.2 Defect Sub-Structure and Phase Composition of the Titanium Surface Layer Fractured When Fatigue Testing ... 174
9.1.3 Structure of Titanium Fractured in Fatigue Tests ... 179
9.1.4 Gradient Structure Developing in Titanium When Fatigued ... 181
9.2 Fracture Surface, Structures, and Phase Composition of Commercially Pure Titanium Disintegrated When Fatigued After Electron-Beam Processing ... 187
9.2.1 Structure of Titanium Irradiated by a Pulsed Electron Beam ... 187
9.2.2 Fracture Surface of Titanium Irradiated by a Pulsed Electron Beam ... 190
9.2.3 Structure Developed in Fatigue Tests of Samples Irradiated by a Pulsed Electron Beam ... 198
References ... 216

Chapter 1
Modifying of Structure-Phase States and Properties of Metals by Concentrated Energy Flows

1.1 Fatigue Failure in Metals and Alloys

To date, engineering, especially the operation of machinery and constructions, faces serious challenges, e.g. the enhancing of materials strength and the extending of their life expectancy, etc. Most high-duty and unique products, machines, and constructions run in conditions of cyclic straining, which will cause fractures even if stressed slightly. It has been suggested that fatigue is a key factor responsible for the rupture of materials under cyclic stressing, i.e. damages tend to accumulate continuously in a material subjected to variable stress, which, in turn, leads to the initiation of a fatigue crack, its propagation, and, finally, to a fracture.

Fatigue failure of machine parts appears to be one of the most frequent reasons for the malfunction of equipment, machinery, and constructions. This fact stems from the peculiarities of high-cycle fatigue. This is the initiation and development of a crack at comparatively low stress, and high sensitivity of fatigue endurance to various factors of design, technology, and maintenance. The third factor is that fatigue resistance varies in an extensive range (dispersion of a lifespan) as compared with static strength. Additionally, cracks are initiated and propagated randomly without any excess displacements till an emergency situation. Therefore, the prevention from fatigue breakdowns of high-duty machine elements (the elongating of their service life) remains a vital issue, especially in industries, where emergencies can bring about fatal consequences.

Currently, the phenomena of fatigue and strength are in focus of research, design, and experimental as well as technological works. Fatigue resistance and fatigue life represent principal criteria when assessing capability and a lifespan of machine components and constructions. In present days, these characteristics are of high relevance for highly stressed and high-duty products, which are operated under cyclic stress of high- and low-cycle fatigue.

Several researchers have argued that the estimation of cyclic strength is a complex issue, since fatigue failure of materials depends on various factors (structure, state of the surface, temperature, and environment of testing, frequency of loading, stress concentration, asymmetry of the cycle, scale factor, etc.). Generally, a fatigue process is related to the moderate accumulation and interaction of defects in the crystal lattice (vacancies, interstitial atoms, dislocations and disclinations, crystal twins, block and grain edges, etc.), which is followed by the initiation and propagation of micro- and macro-cracks.

Such parameters as wear and corrosion resistance as well as the ability to withstand fatigue fracture are essential mechanical characteristics of a material, influencing the reliability of machine components operated under cyclic loads.

During the past 160 years, fatigue of metals has attracted the attention of researchers, although there are still several unsolved problems [1–3]. A number of scientists have argued that a vast amount of accumulated and analyzed information indicates a complex behavior of metals and alloys under fatigue [4–11]. Nevertheless, there is until now no reasonable explanations of damages either under fatigue stress or when diagnosing fatigue. The literature published recently has highlighted the complexity of fatigue, being a result of initiation, accumulation, and interaction of defects in the lattice under fatigue loading.

According to a State Russian Standard (GOST 23207—78), high-cycle fatigue is a state of material when a fatigue rupture or fracture is detected principally under elastic deformation, and low cycle fatigue—elastic-plastic deformation [12]. The other standard (GOST 25.502—79) specifies a lifespan difference in low cycle and high-cycle fatigue ranging up to 5×10^4 cycles. This value is a conditional one and shows an average value of life expectancy when low cycle elastic-plastic deformation turns into high-cycle elastic cyclic deformation. This transformation can be recorded in a range of some hundreds to hundreds of thousands of cycles, being related to cyclic properties of a material, temperature of test conditions, asymmetry degree of a cycle, frequency, and other factors.

The resistance of metals and alloys to high-cycle fatigue depends on several factors, but the main parameter, which is to be taken into account when estimating strength and the life endurance of machines and constructions, is the fatigue curve (or *S–N* curve) (Fig. 1.1) [4], i.e. the ratio between the varying stress and a number of cycles to fracture. Using the fatigue curve, a fatigue limit can be estimated. The latter is the maximal stress a material can withstand until fractures are recorded within an unlimited (>10) number of loading cycles. A number of investigators have pointed out that the plotting of full fatigue curves contributes to the understanding of methods used for the calculation of bearing capacity in all points of a fatigue curve and improves research procedures under non-stationary cyclic loading [6].

Being a material property, a limit state is thought as a full fatigue curve. Typically, a crack is initiated under the surface in all ranges of this curve, from low cycle to high- and ultrahigh-cycle fatigue. In this case, however, a crack appearing on the surface propagates there only for a short period of time. The intensive growth of a crack is documented to be below the surface, i.e. when it has not got through the surface yet. Therefore, it is problematic to control a growing crack—to provide the safe use of construction elements disclinations—with the help of available nondestructive testing means oriented at the detection of through-thickness cracks [13, 14].

1.1 Fatigue Failure in Metals and Alloys

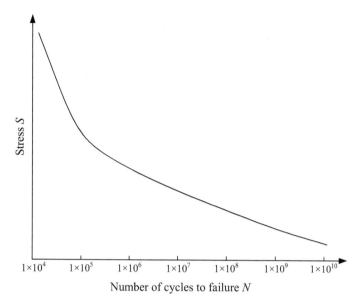

Fig. 1.1 Typical schematic *S–N* curves for metals

A full fatigue curve is divided into two principal parts: low and high-cycle fatigue. Several studies have suggested a conditional line separating these zones is stress equal to a dynamic yield limit [15].

The fatigue of materials includes several periods. Generalized fatigue diagrams demonstrate three stages—gradually accumulating fractures till a fatigue crack is initiated, propagating cracks and breakage gradually accumulating fractures till a fatigue crack is initiated, and propagating cracks and breakage [16].

Numerous investigations [2, 6, 7, 17, 18] have shown that the criteria of resistance to fatigue failure are connected with the structure of material and technology of its manufacturing, machine elements, their construction features, maintenance conditions, and other factors, which can differ significantly even for the identical material.

Central to the entire problem of fatigue failure is the concept of a fracture and the effect of diverse circumstances on a fatigue process. Practically every study on fatigue failure is addressed to this problem. For instance, a fracture of material when fatigue testing is said [18] to be attributed to non-elastic phenomena arising at stress below the elastic limit. It has been outlined that a fatigue micro-crack will be initiated once the microslip has migrated between sliding zones and formed a stress riser. This stress riser forces the development of a micro-crack under cyclic loading.

Continuously growing, it reaches a critical length (Griffith crack [19]), after which a stage of a fragile rupture is registered. Microscopically, the process has been analyzed in the 1960s and later [20–22]. The relevance of external (temperature, state of the surface, and loading conditions) and internal (chemical composition of an alloy, second phase or nonmetallic inclusions, grain dimensions) factors for the structure of a fatigue-worked metal has been determined. Recently, authors have presented evidence that the localization of fatigue cracks is determined by second phase particles, twins, and grain boundaries. One of the findings to emerge from the study [22] has revealed fatigue features are absolutely different at low and high stress.

Several scientists have also suggested certain phases within the accumulation of fatigue failures. (i) Sliding zones are developed. (ii) A surface concentrator is formed. (iii) Ruptures occur on the sliding planes. (iv) Separate blocks of a material move. (v) Cracks appear. The most recent research has also indicated this staging character [23]. In work [24], the facts have been given that fractures accumulate nonlinearly and correlate with a stepwise changing load (increase or decline).

X-ray analysis of normalized steel 35 exposed earlier to fatigue testing [25] has established that plastic deformation tends to be concentrated in narrow zones found almost in all segments of undeformed material, where further propagation of a fatigue crack is recorded.

Therefore, earliest studies have pointed out that a substantial volume of material is undamaged when fatiguing, whereas a huge number of defects are accumulated in some zones, resulting in failures consequently. A number of authors have considered the importance of dimensions of a sample to be used (scale factor) [26].

They have discovered that the bigger a sample, the lower a fatigue limit; in addition, this effect is more determined in high-strength steels. The scale relevance is associated with an increasing surface of a body being deformed, probably, making the origination of dangerous defects more possible. The latter tend to concentrate on the surface and be close to it.

In his study, Yokobori [5] has summarized the data on the kinetics of propagating fatigue cracks and argued a fatigue crack is a stress riser with zones of plastic strain at its tip. This view on the evolution of a crack is identical to cracks of almost all origins [27].

The literatures on mesomechanics of materials published in the last years [13, 24, 28] have contributed to the understanding of fatigue nature. Within this approach the origin of the fatigue fracture is connected with the irreversible motion of material blocks relative to each other and the development of zones with concentrated defects. At this point, the accumulation of fatigue fractures reflects an apparently staging character, well-disclosed via examining a fractal dimension of surface topography [29, 30]. Initially, the authors [31] have paid attention to this fact when investigating into the geometry of spots corroded when fatiguing, and revealed their fractal character.

A distinctive feature of the rupture microtopography in metallic parts subjected to the fatigue is reported to be progression marks oriented perpendicular to the direction of a propagating crack [32]. Progression marks are stripes of consecutive low and high spots or stripes with detachment steps surrounded by these hollows located parallel to a crack front. Once a loading cycle has been completed, a crack (rupture) will move forward at a certain distance. Simultaneously, a row of consecutive stripes remains on a fracture surface. Therefore, stripes indicate a moving crack, normally one step per loading cycle.

Schmitt-Thomas and Klingele have suggested referring to these stripes as progression marks. Every new mark is said to be initiated as a loading cycle is finished [33]. However, in literature there are other theories arguing that the further development of progression marks might result from a series of cycles as well [34, 35].

Progression marks can be continuous and regular (typical for aluminum alloys) with a narrowing gap between them with reduced stress and speeds of crack propagation. They can also be broken and irregular such as those on fracture surfaces of steels. Assessing the space between progression marks, the evaluation of potential material resistance to the propagation of a fatigue crack can be carried out. The smaller the space between marks, the higher resistance to the propagation of a crack material has.

In fact, a huge number of machine parts run in conditions of repeatedly variable loading, so it is hard to overestimate the importance of fatigue fractures. It is the reason for a number of studies which concentrate on the fatigue of metals and have been carried out up to now. Despite a deeply investigated phenomenon of fatigue, there has been little analysis of its nature and it is difficult to forecast the initiation of a crack in such operation conditions.

The science examining the phenomenon of fatigue stands in a logical need of descriptive studies on the evolution of material structure and phase composition. Basically, approaches and models used in mechanics of a deformed solid rely on cyclic loads without considering structural transformations. They are based upon deformation, energy, and power parameters of a stress-strain state as well as criteria relevant for the development of a crack and equations of linear and nonlinear mechanics of the cyclic fracture—these data are required to obtain a key characteristic—a speed of crack propagation.

Nevertheless, it is critical to know the evolution of structure-phase state and dislocation sub-structures to find the regularities of fracture accumulation in the process of fatigue and physical features of this phenomenon at different stages. It is to be taken into account that fatigue cracks tend to be initiated in the subsurface of a material [32]. Therefore, a state of the subsurface influences the fatigue resistance of a material. The pre-hardening of the subsurface can increase cyclic strength, endurance, yield limit, high-cycle fatigue, etc.

1.2 Face Hardening of Metals and Alloys by Concentrated Energy Flows

As mentioned above, a fatigue failure is initiated, as a rule, in the surface of materials. The reason is determined to be the most intensive plastic deformation taking place in subsurface layers, at a depth of about a grain size. The behavior and state of this layer are relevant for endurance till fatigue cracks appear. Being related to characteristics of material on the whole, they influence on the fatigue limit. As a consequence, methods of research into structure transformations on the surface of materials are important to get a deep insight into the fatigue nature.

According to the data of physical mesomechanics, the evolution of a metal state subjected to time-varying till-fracture loading in low cycle and high-cycle fatigue is similar to the behavior of an open system. An open system resists the external cyclic loading in dependence on type, speed, and intensity of inner processes in the surface [36]. A state of metal under the surface is less vital for fatigue resistance, deteriorating proportionally to a distance from the surface. Therefore, serious attempts have been made to slow down the processes of the reaction between the surface of a metal and the environment in order to increase the endurance of metal under cyclic loading. They include the development of layered surfaces and other procedures aimed at modifying surface properties [36].

In this connection, a primary concern of researchers appears to be the expanding of machine parts lifespan and the reliability of machinery, the improvement of their wear resistance by means of diverse face hardening methods. An increase in fatigue strength via face hardening is attributed to two factors: the step-up in surface resistance to plastic deformation and the creation of excess compression stress in the upper layer.

Recent evidence suggests a high-strength specific structure—the "white layer" is formed as a result of an integrated thermo-deformational processing of the surface. In addition, fatigue strength goes up significantly. On the other hand, a weakened subsurface zone is formed due to the thermo-deformational processing in hardened steels; it is a consequence of the second annealing, which deteriorates fatigue strength of a material [37–39].

The use of a certain hardening process is conditioned by specifications of particular products. At present, increased capacities and performance of machinery have raised loads on machine elements, so available methods fail to meet the requirements for their characteristics. Therefore, complex methods of hardening, i.e. a combination of two or more technological processes to provide a better hardening effect than one method only, are in focus of current research [37, 40, 41].

Several studies [38, 39] have highlighted that as a result of a complex method, which involves electro-mechanical processing and the subsequent plastic deformation of the surface, the resistance to fatigue of a treated sample is 30% higher than this parameter of an untreated one. Recently, a principally important issue has remained to be the development of new hardening techniques of machine parts and materials and the updating of the available ones to improve key operational characteristics without affecting the construction and the size of products.

To date, prospective research discusses the development, investigation, improvement, and practical application of face hardening methods based on high-concentrated energy sources, e.g. ion, plasma, laser and electron beams, which further the generation of high-strength nano-dimensional structures in the surface [42–53]. Basic advantages of these modification techniques are as follows: a potential reaction between disperse inclusions and the substrate, the fragmentation of the material structure, the synthesis of nano-dimensional metastable phases as well as nano-composites and intermetallic compounds with unique physical and chemical characteristics.

Several researchers have argued that ion implantation furthers strain hardening. In particular, it is confirmed by the similarity of microstructures of ion-alloyed and strain-hardened materials. For instance, ion implantation, i.e. the consecutive introduction of copper and lead ions into the surface of construction steel 30ХГСН2А enhances its wear resistance and reduces the friction coefficient of treated machine elements operated under sliding friction [53, 54].

Some authors [52] have addressed their investigations to the regularities of wear behavior and fracture of samples with hard layers deposited via ion-laser alloying of austenite steel 08Х18Н10Т. They have demonstrated that hardened layers develop on the surface in the process of ion implantation. Their thickness correlates with the temperature of processing. The findings of friction tests highlight the difference in the wear behavior of samples. It is thought to be related to the thickness of a hard layer.

Another method outlined in recent studies is the face hardening of materials with a plasma arc. Plasma face hardening of a machine part made of steel 40Х13 allows forming of a hardened layer with a depth over 4 mm and martensite structure and equal cross-sectional microhardness [55].

The processing of metallic materials by flows of impulse gaseous plasma leads to the transformation of structure-phase state in the subsurface and formation of a multilayered zone with a diverse structure. A thickness of the modified layer ranging up to 25 μm, depends basically on processing conditions, and is less related to a material [56].

In study [42], authors investigate how the surface and microhardness of low-carbon steel respond to processing conditions of impulse plasma flows. The authors have established that a single treatment involving an increase of energy density hardens a material because of the refinement of formed α-iron crystallites, whereas a multitreatment transforms a low-carbon steel into a high-carbon one.

It is reported that when the first 5–10 impulses are generated, a notable increase in microhardness (by 2–2.5 times) is registered, although a further treatment decreases the hardness of samples slightly [56]. A complex treatment involving several modification methods of the surface gives rise to a more significant hardening effect [57]. Researchers have examined the influence of ion-plasma (TiN) spraying on fatigue strength of steel 20 samples. The authors have shown the maximal fatigue strength corresponds with coating at a phase transition temperature, although other temperatures shorten the total life of samples, making it even worse than of uncoated samples. Employing the methods of acoustic emission, and extended total life of samples with an ion-plasma overlayer deposited at a phase transition temperature has been revealed; that is possible due to the late initiation of a fatigue crack.

Laser hardening technologies are considered to be absolutely prospective methods providing necessary mechanical and tribotechnical properties of machine components [50]. Several studies on wear resistance of laser-processed surfaces have informed that the wear resistance of materials will experience phase transformations when thermal processing is higher than this parameter after traditional volume thermal treatment. The resistance to wear increases due to the more intensive hardening of laser-processed zones and owing to the developed fine-dispersed structure.

Researchers have pointed out the efficiency of laser and laser-ultra-sound methods of treatment for the improvement of wear resistance characteristics under dry sliding friction. In addition, they point at the simultaneous growth of wear resistance and microhardness of surface layers in construction steel 40X.

A complex treatment of steel 20 comprising laser alloying and further nitrogen hardening (LAN-technology) [58, 59] is suggested to generate a single-phase structure of nitrogen-alloyed α-solid solution in a hardened upper layer. Hardness and strength of steel are reported to go up more than twice when processing due to solid solution hardening by nitrogen and dispersive hardening by nitride particles.

An increment of fatigue strength is attributed to a sort and original state of a material, type, and conditions of hardening treatment. The efficiency of the latter is to be estimated to achieve its maximal effect. A method to assess the efficiency of face hardening aimed at the advancing of fatigue strength has been proposed in paper [60] according to the data collected in active tension tests.

A series of investigations carried out to determine how flows of charged particles (electrons, ions), laser and plasma processing, high-frequency impulse currents, electro-mechanical processing, and other energy deposition affect solids have disclosed an immense potential of these methods thought as an instrument to modify surface characteristics of metals and alloys as well as outlined fields of their technological application.

To date, one of the important problems Solid State Physics has been attempting to solve is the establishment of physical mechanisms describing the formation and evolution of structure-phase states and dislocation sub-structure in metals and alloys exposed to external energy sources. It is necessary to experimentally examine the structure and phase composition developing in the cross-section of products when processing. Data of such studies will be relevant for the understanding of the physical nature of transformations, since they allow a purposeful modification of structure and maintenance parameters of a product.

In addition, the necessary complex of high-strength and plasticity characteristics can't be developed unless physical mechanisms and the nature of structure-phase transformations at macro to nano levels are determined.

One of the innovative methods to modify a structure-phase state in the surface of metals and alloys is, as mentioned in Preface, electron-beam processing [43, 45, 46, 61, 62]. This process is characterized by an extraordinary potential to control and adjust the amount of energy supplied to the surface being processed. Within this method, it is possible to expand the zone, where a flow of concentrated energy affects the material. Low coefficients of energy reflection, high concentration of energy in a volume unit of the material; as a consequence, a higher potential to turn the material into a highly unbalanced state distinguish this method from others [57, 63–67].

Comparing with strong ion beams, which are also used to modify the surface of materials, low-energy (<30 keV) dense electron beams are generated with a considerably higher input–output ratio (<90%) in a frequency impulse (to ≈ 10 Hz) mode at lower (an order of magnitude) accelerating voltages without special radiation protection. It is possible because the attendant X-ray radiation is screened by walls of the processing vacuum chamber.

A number of studies have conclusively shown it is better energy efficiency, advanced density of energy in the cross-section of a flow, sufficient repeatability of impulses, and their high frequency that set apart impulse electron beams even from impulse flows of low-temperature plasma if used similarly [57, 67–70].

Electron-beam processing supports ultra-fast heating (up to 10^9 K/s) of the surface to a temperature of melting; the upcoming cooling of a thin subsurface layer (10^{-7}–10^{-6}m) stimulates the formation of maximal permitted heat gradients (to 10^7–10^8 K/m). The latter facilitates the cooling of the subsurface owing to the heat transfer into a depth of material at speeds of 10^7–10^9 K/s. As a consequence, nonequilibrium structure-phase states—amorphous, sub-micro, and nano-crystalline ones can develop in a surface layer.

In their study, Gromov et al. [71] have pointed out a principal strong point of this treatment, i.e. electron energy can be absorbed almost totally, its emission has a volume character, a depth, to which an electron penetrates into the material, as well as the dynamics of heat fields and parameters of a stress wave are broadly variable.

Numerous studies of Chinese colleagues have pointed at a substantial change of structure-phase state in upper layers owing to beams used for modification of alloys and metals; as a consequence, the resistance to wear and corrosion and microhardness increase unlike when traditional processing is applied to the surface [64, 72–75].

A set of samples made of aluminum and cast steel coated preliminary with C, Cr, Ti, TiN powders were processed by electron beams (energy 0–20 keV) using an experimental unit "Nadezhda-2." The experiments have found a significant diffusion effect: during electron-beam processing, diffusing elements of the surface were detected at a depth of several microns in the substrate. A significant increase in wear resistance has also been reported. TEM data have revealed stress waves generated by thermal and mechanical fields, which cause the observed higher diffusion (Fig. 1.2) [76].

Qin et al. [77] have provided details on physical modeling and numerical calculations of thermo-dynamical processes in the material during electron-beam processing. The melting is reported to begin in the surface at a depth of 1–2 mm, with a mechanism of crater developing taken into consideration. Temperature-induced dynamical thermal stress fields can generate three types of stress: quasi-static, thermoelastic, and wave-shaped impact ones. Thermoelastic fields of stress have small amplitudes of 0.1 MPa. The wave of impact stress represents a typical nonlinear wave with an amplitude of hundreds of MPa. Its impact force affects the structure and properties of materials far beyond the heat effect zone. A maximal quasi-static compressive stress in the surface ranges up to several hundreds of MPa and leads to the strain of metal surface.

In conditions of melting, electron-beam processing is reported to enhance hardness and resistance to corrosion thanks to fast crystallization at the speed of 10^7 K/s. A significant result for steels is believed to be the development of nanostructures, which form from a highly overcooled melt. Other driving forces include the purification of

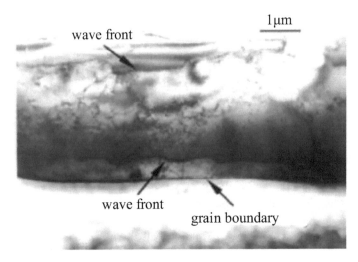

Fig. 1.2 Two stress wavefronts initiating 0.5 mm below the surface and perpendicular to the surface of treatment. The fronts are parallel to the grain boundary, reprinted from ref. [76] (Copyright 2002, with permission from Elsevier Science B.V.)

the melted surface and the strain hardening caused by thermal stress waves. Moreover, metastable phases appear within the treatment. The notion "conditions of melting" doesn't seem to be absolutely correct, since it involves effects of deformation and recrystallization mechanisms. Grosdidier et al. also point at the refinement of grain structure in Fe–Al alloy as well as at hardening, increased diffusion in a solid-state and modification of the grain pattern without changing the surface geometry [78].

Electron-beam processing of carbon steel T8, cast steel D2, and manganese alloy AZ91HP (energy of electrons 27 keV, impulse time 1 μs, energy density 2.2 J/cm^2) initiates significant transformations of structure-phase states and properties of subsurface layers. In addition, a modified layer with enhanced microhardness (several micrometers) is much more expanded than a zone of thermal impact.

Hao et al. [74] have discussed possible mechanisms of crater developing and hardening effect with respect to the depth. In their study, Guan et al. [79] have observed that electron-beam processing of low-carbon steel (0.2 wt% C) in conditions—energy density 1.5–4 J/cm^2 and impulse time 3.5 μs—generates a nano-structure of iron carbide and carbon-oversaturated austenite.

To improve high-temperature resistance to oxidation, Zou et al. [80] carried out a double face treatment of steel H13, which included arc welding and electron-beam processing. Aluminum was sprayed onto steel, a thickness of coating—10 μm; subsequently the material was treated by electron-beams. It has been uncovered in the study that aluminum coating dissolved partly due to high speeds of heating and melting. As a result, a saturated alloyed layer was formed. High-temperature resistance to oxidation enhanced approximately by 40 times in comparison with an untreated sample.

1.2 Face Hardening of Metals and Alloys by Concentrated Energy Flows

Several studies [81–86] have established the refinement of structure-phase states and increase in wear resistance of stainless corrosion resistant steel 316L during electron-beam processing (microsecond impulses). Zou et al. [85] have pointed out that electron irradiation of steel 316L causes partial purification of the surface and the further epitaxial growth in the upper part of the molten layer (2–3 μm), which was softened primarily by tensile stress. The emission of MnS particles played a minor role. Using the methods of electrochemical analysis, the authors have examined the relevance of craters formed because of MnS inclusions for the threefold enhancement of corrosion resistance owing to a uniform layer. The wear resistance is reported to improve significantly in the subsurface layer at a depth of 30 μm with a combined effect of quasi-static thermal stress and waves.

In their next study, Zou et al. [86] have suggested fundamentals describing the evolution of surface craters and the refinement of microstructure as well as those revealing the dominating orientation of crystallographic planes (111). It is possible due to a combined effect of stress fields and temperature after electron-beam treatment. Employing the diffraction methods of back-scattering electrons, intensive plastic strain caused by electron-beam processing has been researched for the same steel 316L.

The effect of electron-beam processing is connected with the orientation. The work has demonstrated, in particular, twinning is activated in grains with an orientation <111> toward a normal line of a sample. High strain gradients of disorientation were developed by the intense crystallographic sliding in other grains. Diverse strain mechanisms and their orientation dependence have been discussed by Zou et al. [84] in terms of two-axial effect of quasi-static thermal stress developed in the process of electron-beam processing.

Electron-beam processing of stainless steel 316L has resulted in purifying, as stated before [85], preventing the formation of craters on MnS inclusions. Additionally, the authors have reported on the homogenization of elements in the molten layer. The combination of these two processes leads to purification. As a consequence, the resistance of steel to pitting corrosion has increased [82].

Electron irradiation of steel M50 (C, 0.75–0.85; Cr, 3.75–4.25; Mo, 4.00–4.50; V, 0.90–1.10 and Fe, wt%) causes softening and worsens its wear resistance because of the increasing concentration of austenite in the molten layer, although corrosion resistance enhances [87].

Xu et al. [87] have argued such factors as the dissolution of carbides and the diffusion of alloying elements contribute to transformations in the microstructure. One-impulse electron-beam processing has brought about a combination of twinning martensite and austenite cells in a molten zone in view of the diffusion of alloying elements around primary carbides. This zone with an excessive concentration of alloying elements represents retained austenite, whereas a zone with a low concentration of alloying elements is transformed in twinning martensite. Once the number of irradiation impulses has been 30, alloying elements will diffuse throughout the molten zone. A molten layer is composed of cellular-type austenite with a diameter of 150 nm. An amorphous structure with a slightly higher concentration of alloying elements has been revealed in a space between austenite grains [88]. In their study,

Zou et al. [89] have investigated nanostructures and metastable phases in the surface of steel AISI D2 treated by electron beams. The features of the surface structure are reported to be nano-dimensional austenite grains (50–150 nm) and dissolving carbides.

G-phase develops from the melt and is observed nearly at room temperature. A martensite transformation is suppressed completely despite a high cooling speed—10^7 K/s. The reason is a more stable austenite phase as a result of the refined grains and chemical modification [89].

The number of impulses is considered to be a critical factor when steel D2 is processed by electron beams: the surface tends to respond differently because of diverse homogeneity and solubility of carbides. Carbides appear to be centers, where craters initiate on the surface. A higher number of impulses leads to the dissolution of most carbides in the surface, so a uniform craterless layer with a significantly stable nano-austenite structure is produced [90, 91].

The quantification of the evolving excess stress in the surface of steel D2 subjected to electron-beam processing has been carried out employing the methods of X-ray analysis. A compression stress in a ferrite matrix is reported to be 560 MPa in the original material made up mainly of ferrite and carbides. A tensile stress reaching a value as high as 730 MPa has been detected in the surface layers as a consequence of electron irradiation. The tensile stress was also found in austenite formed on the surface, varying in a range from 170 MPa (5 impulses) to 700 MPa (25 impulses). Thermal cycling is claimed to be a reason for the evolution of retained stress in the surface layers; since besides melting and phase transformations, it also causes strain after electron-beam processing [92].

A series of studies on titanium alloy TA15 have been carried out using the up-to-date methods of physical materials science (nanohardness, optical microscopy, transmission electron microscopy, etc.). A fine-grained structure is reported to develop in the surface of this material processed by electron beams. Gao [93] has argued it is possible due to the temperature and heat effects of this treatment.

Employing the methods of SEM, TEM, and atomic-force microscopy, several attempts have been made to analyze the electron-beam modifying of titanium alloys TC4 and commercially pure titanium TA2. SEM data have indicated the transformation of a rough surface in the process of electron irradiation. This fact allows assessment of polishing mechanisms, taking into account findings of research into the microstructure of a cross-section. The key factors responsible for enhancing nanohardness are reported to be a fine-grained structure, generating thanks to the rapid crystallization of a molten layer and strain hardening of a subsurface layer [94, 95].

Structure-phase studies on $Ti_6A_{14}V$ treated by electron beams have revealed that a binary phase combination $\alpha + \beta$ turns into a single-phase α'-Ti, and a grain decreases to 100 nm in size. The corrosion resistance of the irradiated layer is claimed to rise significantly as compared with the initial state. Hardness and Young modulus of the surface is lower than in the initial state, whereas the wear intensity of the irradiated layer continues to be constant [96]. Electron-beam processing of α-Ti alloy is reported to bring about the formation of α' martensite in the area of liquid

1.2 Face Hardening of Metals and Alloys by Concentrated Energy Flows

and solid phase transformations. A speed of cooling—approximately 10^7 K/s can be estimated according to former β grains. Aluminum evaporating partly from the surface of a molten layer affects the hardness of the surface and corrosion properties [97].

Zhang et al. have highlighted a product of electron-beam processing of pure titanium is β-phase in the melt zone, which transforms further into fine-grain α' martensite. As a result, corrosion properties enhance and the surface hardness increases by over 60% [97].

The authors have highlighted that the treatment of coarse grain NiTi alloy by electron beams stimulates ultra-fast melting, evaporation, and hardening of the surface. An austenite grain structure B2 600 nm in size was found in the molten layer after crystallization. Zhang et al. [98] have provided compelling evidence that martensite transformation takes place in the subsurface due to the intense thermal stress and the impact waves inducted by electron-beam processing. However, this transformation is unregistered in the upper layer of the melt for the grain structure is undergone the refinement and stabilizes austenite. Another reason for the rising corrosion resistance after electron irradiation is evaporation, which reduces emissions on the surface [98].

If a shape memory alloy Ni–50.6%Ti is exposed to electron beams, ultrafast melting and crystallization of the surface, facilitating the growth of a grain structure of B2 austenite 600 nm in size will be reported. Martensite phase transformation is argued to occur in the subsurface zone owing to high thermal stress caused by electron beams. It is uncovered in upper layers of the molten surface [99].

Gao et al. have established a thin nano-structure MgO layer that forms once magnesium AZ91HP alloy has been treated by electron beams (parameters: energy density 3 J/cm^2, impulse time 1 μs); under this layer, there is a melt at a depth ranging up to 10 μm [100]. Below the molten layer, they have found a zone of thermal impact and a zone of stress waves with stress-induced strain defects. The authors have argued that an intermetallic phase $Mg_{17}Al_{12}$ dissolves almost completely in the melt zone; as a consequence, an oversaturated solid solution forms on the surface. It is the product of rapid hardening. The treatment has significantly enhanced the wear and corrosion resistance of this alloy [100].

Electron-beam processing (energy density 2.5 J/cm^2) has produced a rough surface without visible microcraters in magnesium AZ31 alloy. Selective evaporation of magnesium has brought about the enrichment and hardening of Al surfaces because of the oversaturated solid solution. It is an increase in hardness of the surface that reduces the coefficient of friction by over 20% and improves wear resistance by above six times. Microhardness of samples treated by electron beams is shown to go up enormously reaching a depth up to 0.5 mm, being even deeper than a thermal impact zone (40 μm). The authors have claimed that a propagating impact wave in the process of electron-beam processing is responsible for this phenomenon [101].

Gao et al. have obtained results similar to the abovementioned [101], using alloy AZ91HP in their experiments. In addition to mechanisms discussed in previous research, the authors have emphasized almost complete dissolution of the intermetallic phase $Mg_{17}Al_{12}$ in the process of electron irradiation. Furthermore, the wear resistance of the alloy increases by 2.4 times after the treatment. Face alloying of TiN can also have a positive effect on its wear resistance [102, 103].

In their study, Hao et al. have investigated the microstructure and corrosion resistance of a FeCrAl coating processed by electron beams. They have found porosity reduction in a FeCrAl coating deposited via arc spraying, as well as fewer swelling spots with FeCr columnar grains. The corrosion resistance enhances once electron beam treatment has been carried out with impulses of 50 μs [104]. Researchers have identified that electron-beam processing (impulse time 200 μs) of a FeCrAl coating deposited via arc spraying also brings about the smoothing of the surface [105].

High-temperature resistance to corrosion of the protective layer is tested in saturated solutions Na_2SO_4 and K_2SO_4 at 650 °C. A modified layer was melted through, attaining a depth of 30 μs (energy density 20 J/cm^2). The pattern of Fe and Cr is homogenous in this layer, whereas Al and O formed α-phase Al^2O^3.

High-temperature corrosion tests have demonstrated a shrinkage of around 51 mg/cm^2, which is 20% lower than initially. If energy density is raised, the segmentation of a layer is reported because of big swollen spots and micro-cracks. At 40 J/cm^2, the authors have detected a considerably sharp drop of corrosion resistance [105].

The outcome of electron-beam processing of a nickel-based superalloy DZ4 is a rough topography, since craters tend to appear mainly among dendrites. The thickness of the melted layer is reported to increase slightly with the number of impulses and ranges up to 3 μm at 10 impulses. The corrosion resistance of the modified DZ4 alloy has risen sharply when testing in a 0.5 mol/L solution of sulphuric acid. The study has discussed factors responsible for the improvement of wear resistance, with the surface purification of electron-beam processing taken into account [106].

When processing hard alloy WC/Co by electron beams, Hao et al. have observed the refinement of carbide particles dissolved in binding cobalt. A structure composed of nano-crystalline WC_{1-x} and $Co_3W_8C_4$ phases is formed in the surface layer with a depth of 1 μm. When 20 impulses have been produced in the process of electron irradiation, the microhardness of hard alloy YG6X rises up to 2430 HK and the speed of wear processes is 1/8 slower than in the original material [107].

The processing of WC–6%Co hard alloy by electron beams (energy density 3 J/cm^2 and number of impulses 1–35) has resulted in the melting of WC particles proportionally to the number of impulses; in addition, a plane surface with a net of micro-cracks develops in a layer of several tens of microns. On the other hand, the rapid thermal cycling has initiated the development of a nano grain microstructure WC–6%Co in a layer up to 1 μm in size during electron-beam processing. 20 impulses of electron irradiation have conditioned the generation of a blend consisting of nano grain (20–100 nm) WC_{1-x} and $Co_3W_8C_4$ phases. It provides an increase in microhardness of the surface up to 21.8 GPa and furthers a four fold growth of wear resistance. A perfect irradiated surface has been attributed to the dissolution of solid particles WC and the formation of the amplitude of tensile retained stress due to cracks and local hollows [108].

A group of researchers have carried out several investigations employing the methods of up to date materials science. They have come to the conclusion that

1.2 Face Hardening of Metals and Alloys by Concentrated Energy Flows

impulse-periodic electron-beam processing of industrial steels with a structure-phase states varying from a quasi-homogenous (annealed steel 38ХН3МФА) to highly nonhomogenous (high-annealed steel 13Х11Н2В2МФ), solid alloys with a concentration of a binding agent ranging from 8 wt% (alloy ВК8) to 50% (alloy TiC-NiCr), metallic (alloy Ni-Cr–Al-Y-Ti) and ceramic (composition $ZrO_2 + (6-8)\% \ Y_2O_3$) coatings, as well as powders, furthers the formation of a multilayer structure with significantly refined elements of the structure-phase state in the material subsurface. As a result, the significant improvement of mechanical and tribotechnical parameters of materials is reported. Findings of the study are considered to be fundamentals to develop a technology of electron irradiation of metallic, metal-ceramic, and ceramic materials aimed at enhancing their performance characteristics [109–113].

In their work, Fedun et al. [114] have argued that electron-beam modifying of surfaces of steels М76 and 5ХНМ in two modes (electron energy $E_I = 400$ keV and $E_{II} = 700$ keV, current density $j_I = 30$ A/cm^2 and $j_{II} = 50$ A/cm^2, impulse time τ = 10 μs) produces structures of fine-dispersed martensite, sorbite, troostite and lamellar pearlite [115]. The electron irradiation in both modes causes the melting of surface layers, which constitute a plane boundary with the main zone when crystallizing; the surface displays a rough topography. The authors have also found retained stress of the II grade has the strongest effect on microhardness. A periodic character of microhardness in steels under consideration is attributed to beam fluence. The fluence was 30 J/cm^2 in mode I, and the microhardness decreased steadily. In mode II with the fluence of approximately 30 J/cm^2, it is reported on periodical microhardness.

1.3 The Effect of Electron-Beam Processing on Fatigue Strength of Various Steels

A complex research into the influence of electron-beam processing on fatigue strength of various classes of stainless steel has been conducted by researchers of Tomsk and Novokuznetsk (Russia) scientific centers of Physics of Metals [71, 116–126]. They have established a significant (over 3.5 times) increase in fatigue strength when processing steels 08Х18Н10Т, 20Х13, 20Х23Н18, Э76Ф by low-energy high-current electron beams (Table 1.1).

Utilizing the methods of modern physical materials science, they have carried out a complex analysis of the structure-phase composition dislocation sub-structure as well as fractured surface of steels subjected to electron-beam treatment and found out factors responsible for the growth of fatigue strength.

Fatigue strength of corrosion resistant steel 08Х18Н10Т is reported to depend principally on energy density of an electron beam, being maximal at $E_S = 25$ J/cm^2 (Fig. 1.3). Once the surface of steel has been irradiated by an electron beam, the transformation of the surface layer is registered, i.e. an average grain size drops sharply (1.5–2 times), cells of cellular crystallization develop, particles of the original

Table 1.1 Materials and treatment conditions supporting a maximal extension of fatigue life

Steel	Energy density of the electron beam/(J/cm^2)	The number of cycles left to fracture of untreated steels/× 10^5	The maximal number of cycles left to fracture after electron-beam processing/cycles
08Х18Н10Т	25	1.8	3.5
20Х13	10	1.5	1.4
20Х23Н18	20	1.5	2.1
Э76Ф	20	2.15	2.5

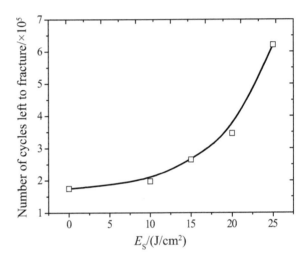

Fig. 1.3 Number of cycles left to fracture in steel 08Х18Н10Т versus energy density of an electron beam

carbide phase dissolve completely, and nano-dimensional particles of the second phase form.

Fatigue strength of corrosion resistant steel 08Х18Н10Т is reported to depend principally on energy density of an electron beam, being maximal at $E_S = 25$ J/cm^2 (Fig. 1.3). Once the surface of steel has been irradiated by an electron beam, the transformation of the surface layer is registered, i.e. an average grain size drops sharply (1.5–2 times), cells of cellular crystallization develop, particles of the original carbide phase dissolve completely, and nano-dimensional particles of the second phase form.

Raising energy density of an electron beam from 15 to 25 J/cm^2, Gromov et al. [71] have reported on the development of a structure with a lower scalar density of dislocations, linear density of micro-twin crystals, and flexural extinction contours. As a result of dropping speed, steels cool and the density of the electron beam rises from 5.7×10^6 to 2×10^6 K/s.

When the surface of steel 08Х18Н10Т has been heated up to pre-melting temperatures at $E_S = 15$ J/cm^2, researchers point out the dynamic crystallization in the surface and the refinement of a grain structure in the steel. Electron irradiation carried out at $E_S = 15$ J/cm^2 is related to micro-twinning processes in the steel as well as to the propagation and sliding of dislocations. 1–2 systems of twin-crystals in a form of parallel rows or sets of parallel plates can be found throughout grains irrespectively to their sizes.

Analyzing the effect of electron irradiation on steel 08X18H10T in their studies, Ivanov et al. [123] have indicated virtually the full dissolution of particles of a primary carbide phase at $E_S = 15$ J/cm^2 and, especially, at energy density of an electron-beam 25 J/cm^2. Secondary particles of titanium carbide TiC (70–100 nm) are argued to form in the steel. A raise in energy density of an electron beam up to 25 J/cm^2 stimulates the development of a structure of dendrite (cellular) crystallization—the size of 320 nm on average.

In the fatigue test of samples treated by electron beams ($E_S = 15$ J/cm^2), their fracture is recorded after $N \approx 2.7 \times 10^5$ loading cycles; this fact shows the fatigue life of steel goes up by approximately 1.5 times. During fatigue, a three-layer structure is formed, consisting of a surface layer of 3–4 μm thick characterized by the absence of deformation relief, a certain transition layer and the bulk of the material. Micro-cracks and micro-pores are detected on the boundary between the surface and the intermediate layer. In the surface there are mesh and cell-mesh sub-structures broken sub-boundaries and shear bands developing either from a mesoband or in a package of micro-twins. The rest of the grain volume represents a band sub-structure [127].

In several studies the facts have been given fatigue tests induce a gradient of structure-phase state of the material, and a majority of criteria change monotonously; however, some parameters tend to vary enormously [124, 125].

Ivanov et al. [123] have compared the structure-phase state of broken samples in the initial steel and those irradiated by a high-intense electron beam. They have come to a conclusion that fatigue life of steel 08X18H10T has enhanced owing to the dissolution of carbide phase particles in the surface irradiated by electron beams; in addition, the refinement of a grain and sub-grain structure has been reported.

In fatigue tests of steel 20X13 as cast and irradiated by electron beams, our group has established a substantial increase in fatigue life of treated steel. Moreover, they have pointed out an explicit correlation between a number of cycles left to fracture and the energy density of an electron beam (Fig. 1.4) [120]. The rates of fatigue life increase are maximal at $E_S = 30$ J/cm^2.

Fatigue loading initiates the development of a structure, which is connected with conditions of preliminary electron-beam processing. When irradiating a layer of $\approx$ 5 μm by an electron beam with $E_S = 10$ J/cm^2 (melting of the surface), the development of several structures has been revealed. They are a band sub-structure, nano-dimensional sub-grains located along the boundaries and in the joints of strain bands and grains, an intermediate layer separating carbide and the matrix, internal fields of stress throughout particles of a carbide phase, and finally, particles of chromium carbide.

When processing a layer of $\approx$ 5 μm by an electron beam with $E_S = 20$ J/cm^2 (melting of the surface), it is reported on the full dissolution of submicron carbide

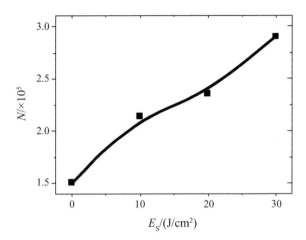

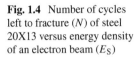 Fig. 1.4 Number of cycles left to fracture (N) of steel 20X13 versus energy density of an electron beam (E_S)

particles (type $M_{23}C_6$) found in as-cast steel, and the development of a grain to sub-grain structure, a mesh dislocation sub-structure, or randomly located dislocations throughout grains.

The irradiation by an electron beam ($E_S = 30$ J/cm², melting of the surface layer (≈ 8–10 μm)) has brought about breakage of crystallization cells, the development of a dislocation sub-structure (mesh, cell-mesh, band ones, broken sub-boundaries, sub-grains), and numerous flexural extinction contours.

Our group has suggested several key reasons for the enhancement of fatigue life of steel 20X13 in the process of electron irradiation. They include the development of a structure of cellular crystallization; the dissolution of submicron globular carbide particles (type $M_{23}C_6$) found in the steel before irradiation; they are considered potentially to be hazardous elements of the structure, which might cause early fractures in the steel during fatigue testing. Another factor is a non-monotonous variation of average grain dimensions. Researchers have indicated the development of a structure with a maximal average size S in a surface layer of steel irradiated by an electron beam (energy density $E_S = 25$ J/cm²), and the formation of a fine-grain structure of the surface when electron-beam processing in pre-melting ($E_S = 10$ J/cm²) and intense melting conditions of the surface ($E_S = 30$ J/cm²).

Fatigue tests of steel 20X23H18 have demonstrated an explicit dependence of fatigue life upon the power density of an electron beam (Fig. 1.5), i.e. the irradiation of the steel surface by an electron beam with a power density 0.4 MW/cm² extends fatigue life more than twice.

Electron-beam treatment and successive fatigue tests give rise to the development of a multilayer structure, which consists of a surface, varying from a few tenths of a micrometer to several micrometers, with an intermediate layer ranging up to 10 μm with a columnar structure, formed in the process of melt crystallization, and a thermal impact layer.

The rupture of steel 20X23H18 under multi-cycle fatigue ($\approx 1.5 \times 10^5$ cycles) is related to a number of factors. The most essential ones include the development

1.3 The Effect of Electron-Beam Processing on Fatigue ...

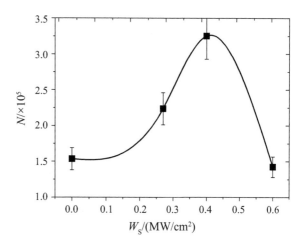

Fig. 1.5 Fatigue life of steel 20X23H18 versus power density of an electron beam

of local areas with a critical structure unable to evolve further, i.e. the exhaustion of plasticity, and the submicron globular particles—stress risers [128, 129]. The upper layer with a critical structure (≈ 10 μm) has a distinct boundary with a layer of the material beneath it. Extended micro-cracks are found in the surface of fractured steel along carbide—matrix boundaries; fragments of steel with a nano-dimensional structure are detected; multiplex micro-twinning is observed and a band sub-structure with fragmental dimensions develops in grains free of micro-twins; in addition, martensite transformation $\gamma \rightarrow \varepsilon$ is registered.

Electron-beam processing of steel 20X23H18 (energy density of an electron beam is varied 20, 30, 40 J/cm^2; an impulse time of an electron beam 50 μs and 150 μs; a number of impulses 3; and pulse repletion rate 0.3) causes the significant (≈ 2.3 times) refinement of a grain structure. The authors suggest a dynamic recrystallization process in the steel, which is initiated by intense stress arising in high-speed cooling [119, 122].

In fatigue testing, zones with a developing critical structure—potential spots of micro-cracks—are formed after more loading cycles due to the processing by high-intense electron beams. Fatigue life of steel 20X23H18 can be increased also by means of other factors initiated by electron-beam processing. They include the refinement of grain structure, the reduction of long-range stress fields, weakening processes of multiplex micro-twinning, and strain transformation $\gamma \rightarrow \varepsilon$ martensite.

Fatigue tests of rail steel Э76Ф have revealed the fatigue life of a material is related to energy density of an electron beam E_S (Fig. 1.6, curve 1). A maximal effect (fatigue life increases by ≈ 2.5 times) is observed at $E_S = 20$ J/cm^2 [117, 126].

When irradiating rail steel by a high-intense electron beam, it is reported on the melting of the surface layer as well as on the development of a structure of cellular crystallization. Gromov et al. [117] have highlighted the carbon-based splitting of the steel surface furthered by a high-intense electron beam and the formation of graphite particles in joints of crystallization cells.

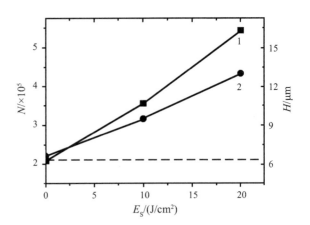

Fig. 1.6 Energy density of an electron beam (E_S) versus number of cycles left to fracture of rail steel Э76Ф (N) (curve 1) and thickness (H) of the surface separated from the base material by micro-pores (curve 2); a fatigue life of as-cast steel (untreated by an electron beam) is given in a dotted line

A net of microcraters covers the surface of steel Э76Ф treated by electron beams at $E_S = 10$ J/cm². Once the energy density of an electron beam has been increased to 30 J/cm², the number of microcraters will reduce, although they expand. In study [117], it has been established that a structure developing in the surface of steel Э76Ф treated by a high-intense electron beam possesses a gradient character. The authors have shown a microheterogeneous structure-phase state develops in the surface of steel (≈ 0.5–1.0 µm) because of high-speed heating and cooling irrespectively to the energy density of an electron beam. This structure-phase state is composed of α-phase grains with crystallization cells, among which nano-dimensional martensite crystals are found, and grains of α-phase with submicron martensite crystals. Besides α-phase retained austenite iron, carbide, and graphite are found.

In steel Э76Ф processed by electron beams (energy density of a beam 20 J/cm²), a layer with pores occupies mainly a boundary between the layer of crystallization and the layer of thermal impact. A layer of crystallization possesses a columnar structure with submicrocrystalline cross dimensions of columns. Therefore, the boundary "hardened layer–base" has a tooth-shaped or needle-shaped profile. The latter fact, as claimed in studies [117, 126], brings about the dispersion of stress risers and facilitates a more uniform plastic flow in the base. As a result, fatigue endurance of steel enhances enormously (≈ 2.5 times).

It has been argued that in steel Э76Ф extended hardened layer with stress risers forms in a layer of ultrahigh-speed crystallization in view of electron irradiation (energy density of an electron beam 30 J/cm²).

In their works, Gromov et al. [117, 126] has established multi-cycle fatigue tests of as-cast steel Э76Ф with pearlite structure have a number of outcomes. To start with, the rupture of iron carbide plates, i.e. they are divided by moving dislocations; iron carbide splits into dislocations for atoms of carbon escape from the crystalline lattice. Secondly, it is the reoccurrence of nano-dimensional iron carbide particles on dislocations (strain aging of steel). Other defects include the development of a sub-grain structure; an increase in a total density of dislocations (dislocations concentrated on boundaries of sub-grains and spread throughout grains); the rise in internal stress amplitude and density of stress risers because of the incompatible deformation of close grains and sub-grains, α-phase and iron carbide inclusions.

To sum up, a high-intense electron beam has a complex effect. Basic reasons for expanding the fatigue life of steels 20X13 and 08X18H10T are discussed to be the refinement of grain and sub-grain structures, and the dissolution of carbide phase particles in the upper layer.

The enhancement of fatigue life of stainless steel 20X23H18 is associated with the refinement of a grain structure, the decreasing level of long-range stress fields, the weakening of multiple micro-twinning processes, and strain transformations $\gamma \rightarrow \varepsilon$ martensite.

Several studies [117, 126] have suggested a principal zone where stress risers appear in rail steel Э76Ф processed by an electron beam is a boundary separating a layer of high-speed crystallization and a thermal impact layer (the bottom of the melt pool). It has been pointed out the fatigue life of steel irradiated by an electron beam improves owing to a needle-shaped profile of the boundary, which, in its turn, causes the dispersion of stress risers and furthers a more uniform plastic flow in the base.

1.4 The Relevance of Face Hardening Methods for the Structure and Properties of Aluminum–Silicon Alloys

To date, aluminum-based alloys have been applied for a variety of purposes in aircraft and motor car industries, in manufacturing of electrical equipment, railway machinery, etc. Recently, researchers have been paying special attention to silumin—a group of alloys based on an Al–Si system. This system is a base for the most present-day aluminum casting alloys thanks to a perfect combination of casting, mechanical, and some special operational features.

Silumin has high corrosion resistance in the humid environment and saltwater; furthermore, it is reported on its better strength and wear resistance in comparison to aluminum. These compounds compete favorably with ferrous metals due to their high-grade physical, mechanical properties, and corrosion resistance. They can replace or force them out of traditional spheres. At present, aluminum-based alloys are the second important construction material, being inferior in output volume to iron alloys only. Therefore, it is an urgent problem to enhance the mechanical and strength properties of silumin.

As seen in the phase diagram (Fig. 1.7), eutectic Al–Si alloys contain 9–13% Si and their structure can be composed of the eutectic and dendrites of α-solid solution and a few primary silicon crystals [130, 131]. Alloys of a system Al–Si, containing iron copper and manganese are thought to be currently prospective light materials. These materials are characterized by outstanding wear resistance, castability, and thermal stability. However, a critical weak point of these alloys is considered to

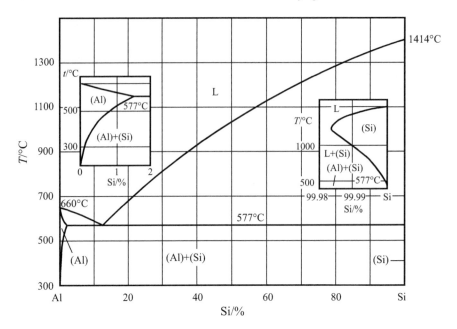

Fig. 1.7 Phase diagram Al–Si, reprinted from ref. [131] (Copyright 2007, with permission from Elsevier Ltd.)

be poor fabricability, since they are sensitive to crack formation. On one hand, iron improves the strength; on the other hand, it stimulates the formation of lamellar intermetallic compounds and reduces the resistance of silumin to cracking. Manganese is revealed to decrease the negative effect of iron, bringing about lamellar or globular morphological transformations of intermetallic compounds, increasing the resistance of the material to cracking and extending its service life, respectively [131, 132].

Silumin compounds are *heavily-deformed* and brittle materials with high-grade specific mechanical properties. Therefore, aircraft, motor car, and other industries are facing a high-priority problem nowadays, i.e. it is necessary to find innovative methods and procedures to improve the structure and plastic properties of silumin. It is a key need to expand its application sphere. Rough crystals (up to 100 μm) of elementary silicon and other excess phases are formed in eutectic silumin, and the eutectic contains lamellar particles of eutectic silicon provided that ingots are crystallized traditionally, using the method of continuous casting [130].

Essential characteristics of aluminum–silicon alloys responsible for their processability and application spheres appear to be casting and mechanical properties [130–132]. Mechanical properties are attributed to the structure and phase composition. They are related to the chemical composition of an alloy, conditions of melting, crystallization, and further thermal treatment. A number of technological processes are available to enhance the properties of silumin. For instance, they can be improved significantly once appropriate technologies of melt processing and thermal treatment

1.4 The Relevance of Face Hardening Methods for the Structure …

have been chosen and a combination of an alloy has been selected correctly [133]. However, it is not always sufficient to comply with all these specifications only to grade up a low quality of silumin.

As temperature has an insignificant influence on the solubility of silicon and thermal treatment has practically no hardening effect on silumin, modifying represents the only method to improve the mechanical properties of aluminum–silicon alloys. The enhancement of a microstructure of alloys is necessary to support their high-degree properties, e.g. grains of the α-phase and silicon are to be more compact; boundaries are to be strengthened by intermetallic modifying phases. These measures are to be taken because rough needle-shaped eutectic and elementary inclusions of silicon cause brittleness, and it worsens as the concentration of silicon goes up. On the other hand, breaking up and spheroidization of silicon crystals in eutectic enhance the ultimate strength by 30–40% and increase the extension coefficient by 2–3 times. Potassium and phosphorous are considered to be the most efficient modifying agents of eutectic [134]. For the same purpose, the authors have used [134, 135] the multi-component modifying of preliminary alloys, which facilitates the rupture of elementary silicon crystals because of activated solid particles in their composition.

A number of researchers [136–143] have carried out systematic investigations into the structure and phase composition of aluminum alloys. To expand their application fields in aircraft, car, and other industries, their structure and plastic characteristics need serious improvement. The authors have claimed that the research and development of alternative prospective methods for the modifying of silumin is a top-priority problem to provide high-grade physical and chemical characteristics of alloys.

Recently, a number of studies have been focused on modification procedures of aluminum-silicon alloys for the manufacturing of high-duty components. They are operated in extreme loading and speed conditions, exposed to a sharp thermal gradient and hazardous environment, e.g. cylinder-piston groups, air compressors of motor vehicles and airspace structures, bearings of turbo-compressors, breaking components in motor car and aircraft industries, pipes for oil extraction branch, etc. These methods use concentrated energy flows and support the development of sufficient physical and chemical characteristics of alloys (laser processing, treatment by plasma flows, strong ion and low-energy electron beams, electric arc melting in the magnetic field, etc.) [144, 145]. The above-mentioned techniques are of significant importance for the modifying of aluminum alloys, which can't find a large-scale industrial application because of low tribological and strength characteristics.

Face hardening can provide an enormous (2–3 times) increase of the ultimate strength, since defects (machining marks, scratches, roughness) on the surface caused by machining are smoothed. The other reasons for the rising strength are compressively retained stress evolving in the surface of a component being hardened and the dispersion of matrix structure and inclusions of secondary phases [130].

In paper [132], the authors have established the role of nano-second pulsed electric currents for crystallization processes and mechanical properties of silumin АК7Ч The crystallization outcomes of silumin exposed to pulsed electric current include the refinement of structure, an increase in ultimate strength by 1.2 times, and the enhancement of flow properties and hardness of the material.

Fast or ultrafast crystallization is reported to be prospective methods for breaking and uniform positioning of silicon crystals in the aluminum matrix. Laser alloying of aluminum compounds by metallic components, their blends and alloys improve performance characteristics of the surface [146, 147].

Laser irradiation of aluminum alloys can initiate essential changes of the structure—the dispersing of structure components, the developing of metastable phases and the defects of crystalline structure, etc.; as a consequence, physical and mechanical properties better [146, 148–151]. Pulsed laser processing of hypoeutectic silumin AK8M3 in the melting mode results in a threefold increase in microhardness in a laser impact zone because structure components disperse significantly. In addition, an oversaturated solid solution of silicon forms in aluminum [152].

The findings to emerge are the melting of a silumin surface and its rapid crystallization owing to the impact of compressed plasma flows. As a result, it is reported on the development of a modified layer (up to 40 μm) with a disperse structure, The findings to emerge are the melting of a silumin surface and its rapid crystallization owing to the impact of compressed plasma flows. As a result, it is reported on the development of a modified layer (up to 40 μm) with a disperse structure, uniformly distributed silicon, and advanced mechanical characteristics. Akamatsu, Wong et al. have emphasized that high-energy plasma flows can essentially reduce the size of silumin structural components, increasing the resistance to wear and strength parameters [153, 154].

Numerous studies have pointed out at a possibility to alter resource properties of a material surface provided that additional levels of submicro and nanoscale structure-phase states are formed in a surface layer. As shown in previous research [71, 116–126], an efficient method of the above modification, bringing about the endurance extension of various steel classes, respectively, is the processing of a material surface by high-intense pulsed sub-millisecond electron beams. The researchers have argued this treatment is advantageous over laser processing as well as plasma flows and strong ion beams.

Electron irradiation makes it possible to modify the structure of a surface layer with a thickness of tenths micrometers; this layer is turned into a multimodal structure-phase state with no serious changes of structure-phase in the main volume of the alloy. A comprehensive analysis of phase composition and defect sub-structure of eutectic silumin (Al–12.8Si) processed by electron beams were carried out by TEM method; parameters of the processing are as follows: energy density 20 J/cm^2, period of impulse impact 50–150 μs, frequency 0.3 s^{-1}, number of electron beam impulses 1–200 [155–157].

The structure of silumin disperses in the process of treatment, coarse inclusions of silicon and intermetallic phases dissolve in a layer ranging to 60 μm with the formation of a cell-dendrite structure. The character and dimensions of its components correlate with the amount of supplied energy. The author has come to a conclusion that the concentration of dissolved aluminum in the melting phase of silicon surface and the subsequent high-speed crystallization contribute to the formation of different structures. In the first place, a structure of cellular crystallization with cells of aluminum-based solid solution. Nano-dimensional fragments of a second phase are detected among cells once a number of irradiation impulses have been set 1 and 5. There are silicon layers on the boundaries of cells. Secondly, aluminum grains appear and contain randomly distributed silicon lamella. Thirdly, aluminum grains are found but without silicon inclusions in their volume. In this case, silicon is reported to be in a form of long layers along grain boundaries. Fourthly, for grains of aluminum, among which globular silicon inclusions are detected, their size varies in a range from 20 to 40 nm [155–157].

As outlined crystallization cells are 160 nm on average during one-impulse processing, an average cross size of silicon layers is 46 nm, and that of silicon particles among the cells is 12 nm. If the number of impulses is raised to 10, an average size of crystallization cells goes up significantly, being 360 nm; whereas a cross size of silicon inclusions increases to 65 on average and that of silicon particles among the cells to 30 nm.

This study provides compelling evidence that a modified layer has two sub-layers. They develop in virtue of the unequal temperature distribution in depth and the difference in melting points that structural components of the alloy under consideration have. The reported structure-phase transformations in the modified layer advance the microhardness and wear resistance of eutectic silumin under the study [155–157].

To date, the potential of electron-beam processing has been in the focus of Chinese researchers in the field of physical materials science. A number of studies [158–168] have discussed that properties of surface layers in eutectic and hypereutectic silumin are modified significantly if affected by short microsecond electron beams using an experimental unit "Nadezhda-2." The authors have established that this effect stems from dynamic stress fields, arising when heating, melting, and cooling. As a consequence, they have observed the considerable refinement of the structure. It is also reported on the improvement of wear and corrosion resistance and an increase of hardness.

Hao et al. [158] have revealed an equiaxial fine-grained structure with a thickness of a few micrometers on the surface of a melted layer when irradiating eutectic alloy Al–12.6Si by electron beams (parameters are as follows: energy density 3 J/cm^2, impulse time 1 μs, number of impulses 10). Under the surface, there is a remelted layer with a thickness of $\approx$ 10 μm containing an oversaturated solid Al solution. The authors have pointed out electron irradiation multiplies the wear resistance (by 2.5 times) owing to the hardening of a fine-grained structure and solid solution.

Gao et al. [160] have demonstrated that in hypereutectic silumin (17.5%Si), high-current pulsed electron beams (energy density 3 J/cm^2, impulse time 1 μs, number of impulses 15) enhance the structure of the alloy containing coarse fragments of silicon.

There is no data on the formation of new phases in the process of electron irradiation, although all diffraction peaks are wider than in the initial state; diffraction peaks of aluminum tend toward bigger angles. In addition, the original alloy with coarse inclusions of silicon has finer fragments after electron-beam processing. It is supported by the mutual diffusion of silicon and aluminum, resulting, therefore, in the formation of a solid solution of aluminum and silicon.

Chemical compositions of aluminum and silicon are arranged in a gradient manner, developing a morphology "halo" in images of the surface. A gradient character of the irradiation zone is also displayed in microhardness patterns of silicon from the center to the edge of "halo" [159].

A new phase has been found in the same alloy irradiated by electron beams; the lattice spacing has decreased significantly as a result of 15 impulses. The measurement of hardness has demonstrated silicon microhardness varies non-monotonously from the center to the edge of the zone, where aluminum and silicon have diffused; in the center of the silicon phase, microhardness drops in a gradient manner as the number of treatment impulses rises.

Initially, the wear resistance goes up but falls with a higher number of impulses; here, the wear loss has been shown to reduce to 84.6% after 15 impulses [161]. Hao et al. [161] have observed in silumin (15% Si) the number of treatment impulses (energy density 2.5 J/cm^2) influences the structure and phase composition in the "halo" zone (Fig. 1.8). This processing furthers the remelting of the surface layer with a thickness of 10 mm (Fig. 1.9).

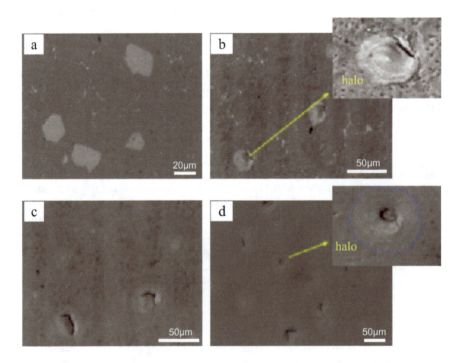

Fig. 1.8 The morphology of the surface in as-cast alloy Al–15%Si (**a**) and treated by electron beams at accelerating voltage 23 kV (number of impulses: (**b**) 5, (**c**) 15, (**d**) 25), reprinted from ref. [161] (Copyright 2011, with permission from Elsevier B.V.)

1.4 The Relevance of Face Hardening Methods for the Structure ...

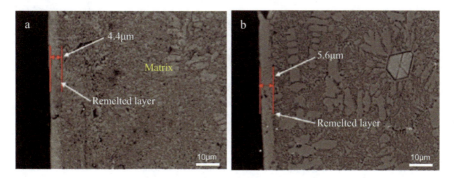

Fig. 1.9 The cross-section morphology of alloy Al–15%Si processed by electron beams (5 impulses (**a**) and 25 impulses (**b**)), reprinted from ref. [161] (Copyright 2011, with permission from Elsevier B.V.)

Hao et al. have also revealed that initial particles of silicon dissolve in the aluminum matrix of hypereutectic alloy Al–20%Si, and an oversaturated solid solution forms in the melted layer in the process of electron irradiation (Fig. 1.10) [163]. The authors [164, 165] have suggested that the diffusion of silicon causes the phenomenon above. X-ray analysis has demonstrated a lattice spacing of aluminum decreases, whereas microdeformation intensifies as a result of electron-beam processing. The data on microhardness of silicon illustrate a gradient manner of changes from the center to the edge of the zone exposed to electron beams.

As a result of electron-beam processing (energy density 1.5–2.5 J/cm^2, impulse time 1.5 µs, number of impulses 15), the microstructure and mechanical properties of alloy Al–Si–Pb are better. An essential increase in wear resistance (Fig. 1.11) of the alloy irradiated by electron beams with energy of 2.0 and 2.5 J/cm^2 has been presented. Utilizing the methods of optical metallography and X-ray photoelectron spectroscopy, An et al. [166] have found out a low speed of wear and friction coefficient at high loads is caused by a greasing film, which covers a surface being tested. The film is a blend of various components of iron, aluminum, silicon, oxygen, and lead in form of a compound Pb_4SiO_6. If the applied load is raised, the authors identify the oxidation wear, the breakage of the film, and the adhesion wear.

An et al. [166] have applied methods of modern physic materials science and pointed out the significant enhancement of a microstructure, wear resistance, and mechanical properties in alloy Al–Pb exposed to electron beams. The development of a greasing film is thought to be a reason for it. The wear behavior varies from adhesion wear at low loads to adhesion wear at high loads. In their study, Hao et al. [167] have indicated the formation of craters and microcraters on the melted surface and a high concentration of vacancies and dislocations in commercially pure aluminum processed by electron beams (energy density 3 J/cm^2) using a unit "Nadezhda-2." The data on microhardness pattern have conclusively shown the modified layer occupies a few hundreds of micrometers and exceeds enormously the heat impact zone. Experimental findings have been compared with prognoses of a model based on

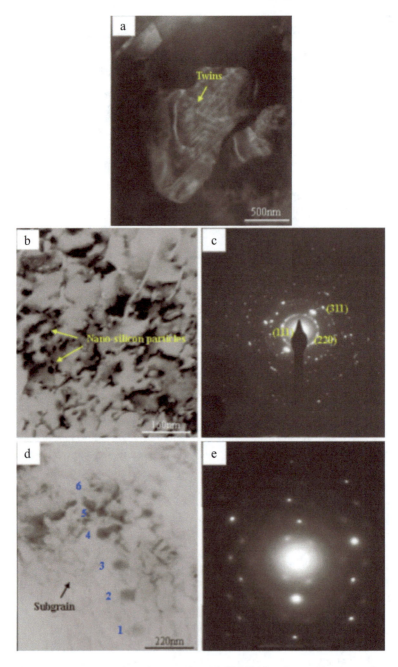

Fig. 1.10 TEM data on the surface of as-cast alloy Al–20%Si (**a**) and treated by electron beams (25 impulses (**b**, **d**)); (**c**)(**e**)—electron-diffraction micro-patterns for (**b**) and (**d**), respectively, reprinted from ref. [162] (Copyright 2010, with permission from Elsevier B.V.)

1.4 The Relevance of Face Hardening Methods for the Structure … 29

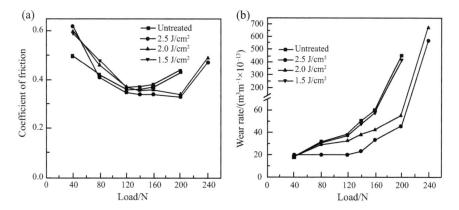

Fig. 1.11 Coefficient of friction (**a**) and wear rate (**b**) versus load for as-cast alloy Al–Si–Pb and treated by an electron beam, reprinted from ref. [165] (Copyright 2005, with permission from Elsevier B.V.)

the numerical solution of heat transfer equations and stress distribution by the finite elements method, and their satisfactory compliance is reported.

Grosdidier et al. [168] have come to a conclusion that electron-beam processing of Fe–40%Al in the heating mode (without the melting of the surface) initiates the formation of a fine-grained structure and modification of a surface texture. It furthers an essential increase in microhardness of the surface in conjunction with the forming vacancies. Nevertheless, corrosion resistance remains stable in this mode because of lacking craters.

To sum up, a considerable amount of literature has been published on the role of concentrated energy flows for mechanical properties of metals and alloys. A number of researchers hold the view that electron-beam processing is one of prospective and approved methods to modify structure-phase states of the surface layers in metals and alloys. Previously conducted studies have suggested that electron irradiation extends the fatigue life (up to 3.5 times) of various steel classes. Investigators have examined the effect of laser, plasma, and electron beams on tribological and mechanical characteristics, the evolution of structure, and phase composition in aluminum alloys, e.g. silumin, although only some studies have attempted to analyze the importance of electron-beam processing for the fatigue life of silumin. Therefore, one of the aims of this work is to clarify the regularities and the physical nature of forming and evolving structure-phase states and defect sub-structure of silumin irradiated by electron beams and subjected to the subsequent high-cycle fatigue loading to fracture.

1.5 Processing of the Surfaces in Titanium and Titanium-Based Alloys

To date, titanium and its alloys are considered to be a prospective construction material. Currently, over 100 types of titanium alloys are available. However, only 30% titanium-based alloys have found a broad application in diverse fields of aircraft and air-rocket industries, shipbuilding, chemical engineering, and medicine. For instance, titanium-based alloys are used in manufacturing of large-dimensional welded and assembled constructions of aircraft, cylinders running under internal pressure in a wide range of temperatures from 196 to 450 °C and a number of other constructs [115, 169–174].

Titanium and its alloys are characterized by the lack of cold brittleness, high plasticity, strength, and corrosion resistance, especially in the oxidizing and chloride environment, as well as low antifriction properties. However, these materials are reported to have low wear resistance, a high pickup tendency, a high coefficient of friction if connected with a majority of materials. The weak points of titanium alloys mentioned above restrict their use in the manufacturing of machine components subjected to friction [172].

At present, engineering has been facing a serious problem, i.e. an efficient method to protect components and products manufactured of diverse materials is to be found. Their surface and subsurface layers tend to fracture in the process of operation; therefore, it is economically efficient to coat high-duty components with protective and hardening layers. The life cycle of coated elements depends upon the composition and quality of the coating, whereas the life expectancy of coatings is related to the method of their production.

Various coatings are deposited onto the surface of products manufactured of titanium and its alloys to improve their properties and protect them from the influence of such factors like temperature, pressure, corrosion, erosion, etc. When the surface has been coated, two vital problems are solved, i.e. original physical and chemical properties of the surface being treated are modified, as well as characteristics, sizes, and mass of the surface undergone to changes in the process of operation are restored.

In literature, coatings are classified into internal and external ones. Internal, also called modified, protective layers have no interface; sizes and weight of a product remain similar. As an internal coating is developed, transformations of a grain structure in the material, the distortion and breakage (amorphization) of a crystal lattice, variations of the chemical composition, and the formation of new phases are reported. External coatings are distinguished by a boundary between the surface of a product and the coating; additionally, the coating heightens the dimensions and the weight of a product. When external overlayers are deposited, properties of a product being treated are modified.

Various methods of the surface processing are available and are used in industry with the purpose to upgrade strength criteria of surfaces, build a structure of a surface and subsurface layers according to workloads. Researchers have highlighted several methods, such as mechanical (polishing, diamond smoothing, etc.),

1.5 Processing of the Surfaces in Titanium and Titanium-Based Alloys

thermal (annealing via induction, laser and electron-arc heating), chemical and thermal (cementation, nitration, carbonitriding, diffusion metallization, etc.), electromagnetic (laser, plasma hardening, ion implantation, magnetic impulse treatment, electron beams, etc.) ones.

Until now, however, there has been no broad classification system of methods to produce overlayers; therefore, they can be systemized from different points of view. For instance, it is reported on internal, external, and intermediate coatings according to the method of their deposition.

External coatings can be formed on the surface of titanium alloys in different ways, i.e. gas-phase deposition, thermal spray coating, gas-denotation, laser cladding, electro spark alloying. Gas-phase deposition as a method of the external coating makes it possible to form functional restoration overlayers. A number of studies have demonstrated that this method to spray coatings onto titanium alloys improves tribological characteristics of a material being processed in spite of pore formation [175–182].

Denotation-plasma spraying represents a prospective method to form biocompatible coatings owing to identical phase composition of a sprayed layer and a base material. Studies [128, 183] have provided evidence of an increase in strength and wear resistance of an overlayer if it is sprayed onto titanium alloys, utilizing the abovementioned method. Despite it, this method suffers from some serious weaknesses, i.e. a non-uniform protective layer and the complexity of the required equipment.

Ghosh [129] has suggested that plasma spraying of powders is a more efficient and wide-spread method. An essential advantage of this method is a possibility to control physical and mechanical characteristics of coating to be produced; in addition, it is a universal one, i.e. merely all materials and their combinations can be sprayed, so multilayered composites are synthesizable. However, plasma-deposited coatings fail to provide a high degree of their cohesion with a base material.

Electro-plasma spraying used to form overlayers on the titanium alloy VT16 enhances mechanical characteristics of the coating [184, 185]. In studies [186, 187], a four–five fold increase in the microhardness of a treated surface has been emphasized. It is a widely held view that laser cladding is efficient to restore old or enhance strength parameters of new machine parts and mechanisms. To date, laser cladding has found a large-scale application in face hardening [188–190] and depositing of cost-intensive coatings with various thicknesses onto titanium alloys [191–193].

The authors [194–197] have demonstrated that thanks to this treatment, the surfaces of titanium alloys with traces of small cracks tend to have better mechanical characteristics similar to those of a sample without defects. However, laser cladding conduces to the formation of a groove in the zone of treatment, bringing about disorders of geometry in original samples. In addition, a plating layer of a powder appears on the surface, being burnt then by a laser ray.

Interestingly, titanium carbide is used in the most listed above methods of surface treatment. It is primarily because of its sufficient hardness and wear resistance at low and high temperatures, high melting temperature and low coefficient of friction, adhesion properties, reliable cohesion with the surface to be covered (titanium alloys), chemical stability in the aggressive environment and resistance to slagging.

Another critical factor contributing to the broad application of titanium carbide is its low price and availability. Heighten concentration of titanium carbide in the alloy produces a beneficial effect on its wear resistance. Protective layers significantly expand a lifespan of a machine part and a machine on the whole. In dry friction tests, the wear resistance of aluminum and titanium alloys coated with protective titanium-based layers via laser surfacing is reported to be much higher than in similar uncoated alloys.

In the first place, titanium and its alloys can be saturated with gases, being in a hot state. Nitration, boriding, oxidation, and cementation are processes used in industry to harden titanium alloys. A long-time treatment at high temperatures can provide a maximum thickness of the hardened layer. Such methods of chemical and thermal treatment advance substantially antifriction characteristics of titanium alloys, e.g. better surface hardness, resistance to wear and corrosion. The use of other traditional chemical and thermal processing methods for hardening of the surface fails to be efficient since titanium alloys have the particular affinity with nitrogen and oxygen.

In the process of chemical and physical deposition of an internal coating onto titanium alloy a titanium diboride TiB_2 and borides TiB_2 and TiB are synthesized on the surface in the zone of alloying, which is 35–40 μm in size [198]. Carbonization of titanium alloys is carried out by open-air methods. This treatment of a titanium base reduces the strength of a coating because of titanium oxides and borides forming on the surface.

In studies [199, 200], it has been stated that a continuous layer of titanium carbide can't be formed during the deposition of internal coatings. This fact is a serious limitation to provide a maximally possible degree of face hardening. In some works, it has been suggested that carbide layers generated when diffusion surface plating with chromium or titanium contribute to the formation of a dense layer with high corrosion resistance and physical and mechanical characteristics. This effect is possible due to carbon.

Chemical and thermal treatment of inner surfaces by means of nitration and oxidation suffers from other disadvantages, i.e. low efficiency, high power consumption, and a limited depth of a layer being hardened. The raise of temperature expands a grain in the product to be processed, causes the diffusion of hydrogen, deteriorating plasticity and viscosity as a consequence.

To sum up, it is absolutely indispensible to search for innovative less time and power-consuming technologies to produce an overlayer with outstanding tribological parameters. The analysis of currently available methods to strengthen surfaces in diverse products has revealed that the use of concentrated energy flows is a prospective one. It is believed, a mainstream direction in the search for innovative processing methods is the acceleration of heating, cooling, and deformation. The purpose is to make defects of crystal lattice in the processed surface more concentrated, as well as to evolve their pattern, improving, as a consequence, mechanical, and other essential characteristics of the material. The discussion of concentrated energy flows is focused principally on a target effect on the treated surface, produced by an energy flow with a thermal density of 10^3 V/cm^2.

In practice, laser irradiation, electron beams, and plasma flows are principal methods of face hardening [201]. All methods of face hardening listed above have both advantages and disadvantages. A major drawback is the valuable equipment for their realization, as a consequence, a cost-intensive treatment. A compressed electric arc may be used to reduce cost price during the processing of different surfaces.

In works [202–204], the authors present findings of research into carbonization of diverse metals when surface alloying, i.e. introduction of carbon-graphite fibers via electrical explosion. The authors have shown that electrical explosion carbonization of titanium and its alloys turns a phase composition in a layer to be modified into a solid solution of carbon in titanium; additionally, isolated particles of titanium carbide and free carbon penetrate through the zone of melting and alloying. A supplementary treatment of the surface is required to dissolve graphite and raise the concentration of titanium carbide. It is efficient to process the surface subjected to electrical explosion alloying by high-current pulsed electron beams.

Researchers [205–207] have tested this supplementary electron-beam processing for additional face hardening of commercially pure titanium VT1-0 and constructional carbon steel 45 subjected previously to electrical explosion alloying. The abovementioned studies were conducted to deposit internal coatings of titanium carbide when dissolving particles of carbon–graphite fibers introduced into the zone of electrical explosion carbonization.

1.6 The Use of Concentrated Energy Flows for the Face Hardening of Titanium and Its Alloys

The hardening of metals and alloys via irradiation of the surface by concentrated energy flows is the heating of a surface layer being treated up to the temperature of melting or above it if an absorbed power density q is increased. In case an absorbed power density amounts to GV/m^2, a source of heat can be called concentrated [208]. The processing is carried out in the impulse mode in order to achieve a needed q. Sometimes, i.e. in electrical explosion alloying, the treatment is combined with a mechanical impact; pressure on a surface to be treated may range up to 10^2–10^3 MPa. Once a source has stopped supplying the power, the surface cools at a high speed. This process is associated with the formation of annealing unbalanced structures with a high concentration of defects in a crystal structure.

The strength of the surface tends to go up because an amount of absorbed power density increases, the time of the impulse impact is reduced. A short-time impulse impact (time τ varies in a range from 10^{-6} to 10^{-3} s) makes less possible a risk that micro-cracks appear and propagate in the surface, worsening tribotechnical characteristics of the modified surface [112]. This study has linked serious changes in mechanical, physical, and chemical properties of the surface with a treatment, which combines melting with the further alloying of the melt and self-annealing. New hardening phases (carbides, borides, nitrides, etc.) develop in a layer with a thickness up to 1 mm. Surface alloying of construction and tool steels and alloys by concentrated energy flows has both its positive and negative sides. Therefore, recent research has been focused on the complex surface treatment combining different techniques, which intensify the effects of each other.

In recent studies [133, 209], researchers have investigated the saturating of the surface in titanium alloy VT9 when laser face alloying by nickel, chromium, iron, and manganese. The use of copper facilitates the overheating in the center of a spot and the developing of a copper layer on a titanium alloy. As a result, this layer is distinguished by relatively low microhardness. Chromium, iron, and nickel are reported to be the most efficient saturating elements, since the microhardness of the modified surface ranges 9500–10,500 MPa. When alloyed by manganese, the microhardness of a titanium alloy is estimated to be 8700 MPa.

The analysis of diffraction patterns has revealed that reflexes TiO_2 found in the initial material disappear due to laser alloying of a titanium compound. It is evident the reaction of VT9 with oxygen of air tends to weaken at high-speed thermal treatment as a consequence of a considerably shortened heating period of the metal to high temperatures. The nitration of titanium alloy VT16 in plasma of semi-self-sustained low-pressure arc discharge in a mixture of nitrogen and argon has a positive effect on adhesion wear. The most relevant findings of research into the effect of a full cathode under high-temperature ion nitration in a glow discharge on constructional and tool steels have been pointed out in recent studies [210, 211]. In this case, the thickness of the hardened titanium alloy VT6 layer is 3–5 times higher than in the process of nitration in a system iron-nitrogen. The data on surface microstructure indicates no coarse grain structure, and reports that the microhardness after the processing is 4.1 times higher than before. A number of scientists are engaged in the nitration of titanium alloy VT1-0 at low pressure in plasma of a glow discharge with a hollow cathode [212–214].

Recently, a method of electrical explosion processing has been developed intensively. The investigation into the surface alloying by impulse multiphase jets of products generated in the electrical explosion of conductors broadens the scope of hardening of various surfaces. Electrical explosion alloying has an impressive effect on microhardness, heat and wear resistance, and other important characteristics of the surface in machine parts, constructions, and tools. Heat and pressure characteristics of this treatment enable obtaining structure and phase states of the surface, which can't be provided by other types of hardening.

At present, electrical explosion alloying is thought to be a prospective technique. This is caused purely by a relatively low cost of equipment needed for this process. Workshops where this equipment is installed are to meet no special requirements. In addition, this equipment is compact, allowing its introducing into the low-scale manufacturing process of machine components or applying in small businesses [215].

The influence of an ultra-sound plasma impulse jet the base material in electrical explosion alloying on has been suggested [194, 203]; additionally, this jet appears to be a heat source for the alloying of surface layers. In the process of treatment close to the surface, an enormous increment of pressure is detected. A specific power density q and impact time τ are thought to be basic physical criteria of alloying, which determine a thermal field in the material and a degree of alloying. If a specific power density of an electron beam ranges to 1 kV/cm^2, a surface being irradiated reaches a temperature of melting. Here, a shocked layer is formed close to the surface with a temperature of 10,000 K and pressure to 10^6–10^7 Pa. In the zone of melting the surface is saturated with components of plasma; alloying elements penetrate throughout the melt, remixing in a convective manner with the base material.

As a beam of electrons has a short impact time on a surface to be processed in metals and alloys, the alloying of surfaces reaches a depth of 20–40 μm, an area of 10–15 cm^2 in size. The most valuable effect of alloying is achieved provided that treatment is carried out in impulse mode with an impact time of 10^{-6}–10^{-3} s. A study on the reaction of multiphase jets with metals taking place in electrical explosion alloying has pointed out that both plasma components of a jet and condensed particles facilitate the alloying.

Commercially pure metals and compounds are used as a conductor to be exploded for the modifying of structure-phase states and developing of certain process properties in the surfaces of various metals and alloys via electrical explosion alloying. A number of substances with high electric conductivity, e.g. carbon–graphite fibers, are applicable as a material to be exploded. Weighted powder portions of different substances and compounds placed into the zone of explosion are a source of alloying additives to saturate a surface to be treated. Frequently, weighted powder portions of boron, carbon, borides, carbides, oxides, and nitrides are used in electrical explosion alloying. These alloying elements, being partly in a plasma state, are entrained and transferred onto the irradiated surface by a plasma jet.

Electrical explosion alloying has a low capability and combines a local heat impact onto the surface and its saturation with alloying elements. The latter is selected from a wide range of exploded conductors and weighted powder portions of diverse substances such as carbides, borides, etc.

Despite certain advantages of electrical explosion alloying by concentrated energy flows over other face hardening methods of metals and alloys, its usage faces some restrictions. It is related to the topography and uncompleted structure-phase transformations in the impact zone of a structurally heterogeneous plasma jet. To date, there is little data on features and mechanisms explaining the formation of an impulse plasma jet and its influence on the structure and properties of the treated surface.

Recent research has demonstrated that the opportunities of electrical explosion alloying may be expanded by means of the repeated electron-beam treatment of the surface to be alloyed. Electron-beam treatment is conducted using low-energy high-current electron beams. A double treatment levels roughness of the surface, decreases a gradient of microhardness over the depth, lowers stress on the contact boundary between the surface and base material, and reduces a coefficient of friction as well as raises three–five fold microhardness and wear resistance of the treated surface.

1.7 The Modifying of Structure and Properties in a Complex Surface Treatment

An impulse plasma jet is generated in electrical explosion alloying due to the discharge of condenser battery onto a plasma accelerator; a conductor to be exploded is used as its material. A plasma impulse jet is both a source of the thermal impact and an instrument to treat the surface. As mentioned above, a pressure at the surface to be treated rises significantly in electrical explosion alloying.

A group of authors [145, 216] have argued that a product of electrical explosion of conductors is a multiphase system composed of explosion products. When a high-speed plasma jet is formed, products of electric explosion reach the surface of treatment at different points of time. Particles of the powder melted on the surface as well as condensed drops of explosion products are at the back of a jet, reaching a surface to be processed at the moment it melts or crystallizes. So, surface treatment of metals and alloys by high-concentrated energy flows stimulates the development of diverse topography.

An impulse flow affects the surface of metals and alloys, causing its melting and non-uniform radial spreading of metal on the surface from the center to the periphery. In a number of works [188, 217–219], it is reported that the spreading of metal is less obvious provided that weighted portions of powders are used in electrical explosion alloying.

After electrical explosion alloying, the topography is wavy; diverse defects such as craters, overlaps, hollows, drops, bubbles, pores, and micro-cracks are detected [200, 202, 216, 220–222]. This fact is related to the impulse impact and un-uniform distribution of alloying elements during spraying and melting as well as to the boiling and evaporating of the upper layer in the surface of treatment.

Analyzing the reaction of strong impulse flows with materials [223], the authors have revealed fewer serious fractures in the surface of ferrite steels than in austenite ones. There are fewer defects such as pores, drops, solid melt, and spots of precipitated copper on the surface of ferrite steels. Besides defects mentioned above, on the surface of austenite steel there are also craters, bubbles, overlaps, and traces of opened bubbles. The examination has pointed out that gaseous microinclusions play an important role in the development of surface structural defects in the melt of a surface layer. They tend to be formed from ions implanted into the material, atoms of inflating gas, and volatile components, which constitute a material to be irradiated [224, 225].

Yun et al. [219] have investigated the topography of the surface zone being alloyed, which evolves in electrical explosion of metal foils and come to a conclusion a well-developed topography appears on the surface being processed irrespectively to the base metal of a conductor to be exploded. For instance, relatively level zones dominate on the processed surface of a system Fe–Al, whereas shares of well-developed and smooth topography are virtually identical in systems Fe–Cu and Ni–Cu.

In a triple system Fe–Al+B, a correlation of smooth zones and areas with well-developed topography is reported to be 1.5:1, in a system Fe–Cu+B—1:5, and in a system Ni–Cu+B—1:8. When alloying the surface of steel 45, similar mechanisms and final surfaces with higher roughness are detected [216, 226, 227]. It is one of the characteristics typical for the development of the surface topography during electrical explosion alloying of metals and alloys [58, 228, 229]. The additional machining is needed for the improvement of the surface topography and its tribological properties.

So far, a few studies have suggested the effects of electrical explosion alloying may be intensified by means of the subsequent processing by high-current electron beams [219, 228]. The authors [112] have found out the original roughness of a surface being treated can be straightened out significantly if processed by high-current electron beams.

A number of investigations [230–232] have been focused on regularities dominating the formation of structure-phase states as well as on the enhancement of microhardness and wear resistance of metals and alloys in the process of combined treatment. The topography of surfaces processed by electron beams after electrical explosion alloying is smooth, no micro-cracks, and micro-pores are found.

The zone of a combined treatment has a gradient structure. Throughout it three layers are detected: a subsurface layer, an intermediate layer formed as a result of electrical explosion alloying, a thermal impact layer, the hardening of which is possible thanks to structure-phase changes in the alloy base.

A number of impulses and a time of the electron beam impact are relevant for the thickness of the treated surface. For instance, the depth of the impact zone under the carbonization of titanium alloys approximates to 70 μm, and it is 60 μm in aluminizing, and even 90 μm in boron-aluminizing [230, 231]. Under combined treatment, the multiphase structure develops, making the surface harder as a consequence.

There is a huge agglomerate of experimental data on the combined treatment of various alloys and metals [145, 230, 231]. Experimental results have shown that the wear resistance of the surface layer is three–seven times higher, and the microhardness is 10 times higher as a consequence of the combined treatment.

In electron-beam processing of commercially pure titanium, its microhardness is 12 times higher for the carbonized surface, three times higher for the aluminized surface and five times higher for the boron-aluminized layer, respectively. In a layer of titanium VT1-0 subjected to electrical explosion alloying, there are isolated particles of titanium carbide with a size of 1–2 μm as well as particles of carbon–graphite fibers, which were involved into the impact zone of impulse treatment and failed to dissolve (Fig. 1.12a).

In the process of the subsequent electron irradiation particles of carbon fibers dissolve close to the zone of alloying. So, a continuous layer of titanium carbide with a thickness of 20–25 μm is developed. Particles of carbon-graphite fibers were detected at the depth of 60 μm. When chemical etching a dendrite structure is detected in these sub-layers, this structure is composed of titanium carbide in a metallic matrix (Fig. 1.12b). Microhardness in the surface layer ranges up to 2400 HV, in a zone of a dendrite structure to 600–1000 HV. Insignificant hardening (up to 300 HV) is discovered in the zone of thermal impact [233]. Models describing the dissolution of carbon particles clarify these phenomena: coal-graphite particles consist of tiny

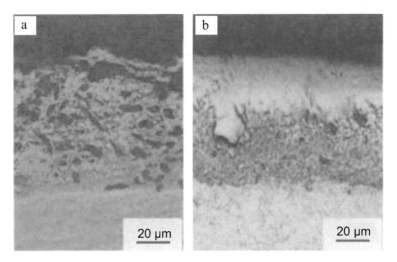

Fig. 1.12 Structure of the surface layer in titanium VT1-0: electrical explosion alloying (**a**); electrical explosion alloying with the subsequent electron-beam treatment (**b**)

microfibrillae (10 nm), liquid titanium flows among microfibrillae if a surface of metal is exposed to an electron beam [233].

Examining the surface of titanium alloy VT6 irradiated by an impulse plasma jet, researchers have also found fragments of coal–graphite fibers. In their study [233], Mikheev et al. have provided a deep insight into the complex dissolution kinetics of carbon-graphite fibers, which penetrate into the surface layer under electrical explosion alloying. Once electron-beam processing has been completed, coal–graphite fibers will dissolve completely in the intermediate layer, occupying a space between melted and solid layers. The time of dissolution is reported to be shorter than stated in the theory of fibers. As a result of treatment, structures of cellular and dendrite crystallization are found on the surface.

A group of researchers have been investigating the electrical explosion aluminizing and boron-aluminizing of titanium alloys by an electron beam [229–232]. It has been reported on a protective layer, consisting of microproducts generated in the explosion of aluminum foil and boron and depositing at the back of a multiphase plasma jet. Micro-cracks and micro-pores are detected on the surface. The subsequent processing by electron beams is related to the high-speed melting and cooling of the surface as well as to its smoothing. It is possible because of capillary forces and fewer micro-cracks.

A gradient structure of the aluminized zone to be generated in the process of electron-beam processing is not similar to that formed in electrical explosion alloying. There is no thin surface nano-composite layer in it, furthermore, no layer of cellular crystallization is found, and in addition, no thin nano-structure layer can be identified on the contact boundary between the alloyed zone and the base. It might be caused

1.7 The Modifying of Structure and Properties in a Complex ... 39

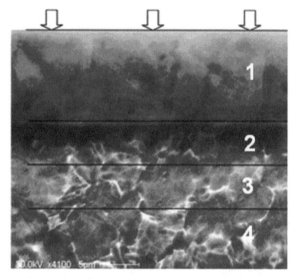

Fig. 1.13 A cross micro-section of commercially pure titanium VT1-0 sample exposed to electro-alloying aluminizing and electron irradiation

1 — subsurface layer; 2 — intermediate layer;
3 — layer of thermal impact; 4 — material base

by the evolution of structure-phase state in the surface of the alloyed zone in the process of supplementary electron irradiation.

Through investigating into cross micro-sections, a lamellar structure with three layers (subsurface, intermediate, and layer of thermal impact) is detected (Fig. 1.13). The arrows in Fig. 1.13 show the hardened surface. The subsurface differs from the rest of materials by the contrasting corrosion; its thickness varies from 15 to 50–60 μm.

A feature of a subsurface layer is its different thickness, there are mainly spherical as well as lamellar (needle-shaped) and roundish (globular) structures in this layer. A length of lamella varies from 3 to 7 μm, and their cross-sectional dimensions are in a range of 0.2–0.3 μm. The lamellas are grouped in packages of 3 to 5 items. An intermediate layer of 1–5 μm has a grain structure; sizes of grains vary from 1 to 2.5 μm. Average dimensions of grains increase as far as from the treated surface in the thermal impact zone. Second phases are found on the boundaries of grains.

A gradient structure of the boron-aluminized zone in titanium processed by electron beams is similar to that of electrical explosion alloying. On the surface there is a nano-composite layer, consisting primarily of aluminum and titanium borides, at the depth layers with cellular and grain crystallization structures are detected. However, junctions of layers are less marked, and a boundary between the alloying zone and the base is indefinite.

Despite the intense effect on a surface to be alloyed in the process of electron irradiation, the structure of all layers is submicrocrystalline due to the impulse-periodic character of electron-beam processing [216, 230, 231]. Interestingly, the subsequent electron irradiation carried out in optimal conditions increases the microhardness of the hardened layer (by 2.5 times, up to 40 μm) in comparison to a sample treated in electrical explosion alloying; microhardness of the surface layer remains similar.

Electrical explosion boron-aluminizing enhances the microhardness of the surface by three times, and the further electron-beam processing results in its 5.5 fold increase (up to 1800 HV), a total depth of hardening is reported to be 60 μm.

The use of a weighted portion of titanium diboride powder is suggested to provide the strongest hardening effect both as a result of electrical explosion alloying (titanium diboride, silicon carbide, and zirconium oxide) and electron-beam processing [234].

The physical nature of hardening in the combined treatment may be revealed in X-ray diffraction analysis and electron-microscopic studies. According to X-ray diffraction findings, a principal phase in the modified layer is α-Ti in samples exposed to electrical explosion carbonization by titanium diboride powder and irradiated by electron beams. As a result of electrical explosion alying, its volumetric share is 17%, rising from 50 to 72% after electron-beam processing thanks to a going up energy density of an electron beam. New formed phases are reported to be TiC, TiB_2, Ti_3B_4, C, and B_8C. In all conditions of electron irradiation, a principal additional phase appears to be titanium carbide, and its volumetric share is 60% after electrical explosion alloying and 40% as a consequence of electron-beam processing (energy density of an electron beam 60 J/cm^2). In electrical explosion carbonization of titanium alloys with weighted powder portions of titanium diboride, silicon carbide, and zirconium oxide, researchers have identified a hardened zone with varying thickness, element composition, and structure-phase state. The highest microhardness is registered when titanium diboride powder and commercially pure titanium VT1-0 are used as the base: the surface microhardness is 14 times higher with a total thickness of the hardened zone of 65 μm. Electron irradiation allows developing of a hardened zone with low surface roughness and well-distributed alloying elements. After the completion of the subsequent electron-beam processing, a thickness in the hardened zone is higher with a low general level of microhardness. A thickness of the hardened zone is reported to be higher, up to 100 μm if titanium diboride is used. Volumetric maximums of microhardness are found in the depth of the hardened zone. The resistance to wear of the surface increases by 8.2 times, and a coefficient of friction falls approximately by 1.2 times as compared with the original material. Two layers with dendrite structure of diverse dispersion are identified throughout the zone of hardening [133, 155, 219, 220, 230, 234].

X-ray dispersion studies have revealed that α-Ti (17%) is a basic phase in the zone of electrical explosion alloying by TiB_2 powder; as a result of treatment, its share increases with relation to a rising energy density of an electron beam. New-developed phases include TiC (60%), TiB_2, Ti_2B_5, Ti_3B_4, TiB, C, and B_8C. In all conditions of electron irradiation, the principal additional phase is reported to be TiC (40%); energy density of electron beam is set to be 60 J/cm^2. In the hardening zone, a multiphase sub-micro to nano-dimensional dendrite structure has been revealed. TiC inclusions contain practically no defects; a dislocation sub-structure is found in grains of α-titanium, and particles of hardening phases are also detected (SiC, TiSi, and $TiSi_2$ when using silicon carbide, TiC, and ZrO for zirconium oxide, respectively).

The outcome of carbonization is a maximum microhardness of 800 HV on the treated surface, dropping in a monotonous manner up to 180 HV at a depth of approximately 50 μm. Its boost may be caused by hardening particles of titanium carbide formed in the zone of alloying [229]. As electron-beam processing is finished, a maximum of microhardness (2500–3000 HV) is recorded not on the surface but at a depth of approximately 20 μm. Microhardness is measured and is 14 times higher than that of base material. A raise in impulse time from 100 to 200 μs leads to a slight increase of microhardness close to the irradiated surface and a second maximum at a depth of 70–80 μm. Such a pattern of microhardness points at an intense reaction of titanium with carbon both in the zone of remelting and in deeper layers in the process of electron irradiation as well as at the role of mechanical stress new-developed layers cause each other. In electrical explosion alloying of commercially pure iron and nickel, microhardness is reported to be five times higher as a consequence of single-component alloying, and 14 times higher for complex alloying.

According to data of micro-diffraction TEM-analysis, such phases as titanium carbide (TiC), α-titanium, nano-dimensional graphite with a cubic and hexagonal crystal lattice, titanium dioxide (α-TiO$_2$) are discovered in the zone of alloying in titanium compounds.

It is suggested that a grain structure of titanium has dislocations with a scalar density of approximately 5.6×10^{10} cm^{-2}. In a number of studies, authors have revealed that globular titanium carbide particles (4–6 nm) can appear on the surface and in the subsurface layer of carbon fibers as well as on the surface and in the subsurface of titanium grains [228, 231, 234].

A complex treatment combining electrical explosion alloying and electron-beam processing expands the alloying zone, levels the roughness of the surface, decreases a gradient of microhardness over the depth and internal stress on its boundary with the base material, heightens three–five times microhardness and wear resistance of the hardened surface, and a coefficient of friction declines as well.

This treatment is reported to support full realization of all hardening mechanisms. If processing parameters of hardening specified according to certain aims and objectives, an optimal state of the structure can be developed to provide a maximal level of construction strength. The latter is possible due to the solid solution hardening with alloying elements, a high density of dislocations, and the development of a super fine-grained structure, which forms at high cooling speeds, as well as the second phase particles.

References

1. Cui, W.: A state-of-the-art review on fatigue life prediction methods for metal structures. J. Mar. Sci. Technol. **7**, 43–56 (2002)
2. Stephens, R.I., Fatemi, A., Stephens, R.R., Fuchs, H.O.: Metal Fatigue in Engineering (2000)
3. Schutz, W.: A history of fatigue. Eng. Fract. Mech. **54**, 263–300 (1996)

4. Bhaduri, A.: Fatigue. In: Mechanical Properties and Working of Metals and Alloys. Springer Series in Materials Science, vol. 264. Springer, Singapore (2018)
5. Yokobori, T.: Physics of Strength and Plasticity. MIT Press, Boston (1969)
6. Kennedy, A.J.: Processes of Creep and Fatigue in Metals (1962)
7. Golovin, S.A., Puskar, A.: Microplasticity of a low-carbon steel at low and high loading frequencies. Kov. Mater. **16**, 426–437 (1978)
8. Collacott, R.A.: Mechanical Fault Diagnosis and Condition Monitoring (1977)
9. Pisarenko, G.S., Troshchenko, V.T., Strizhalo, V.A., Zinchenko, A.I.: Low-cycle fatigue and cyclic creep of metals. Fatigue Fract. Eng. Mater. Struct. **3**, 305–313 (1980)
10. Krautkrämer, J., Krautkrämer, H.: Ultrasonic Testing of Materials. Springer-Verlag, Berlin Heidelberg (1990)
11. Ivanova, V.S., Goritskii, V.M., Orlov, L.G., Terent'ev, V.F.: Dislocational structure of iron at the tip of a fatigue crack. Strength Mater. **7**, 1312–1317 (1975)
12. Miller, K.J.: Metal fatigue—past, current and future. Proc. Inst. Mech. Eng. Part C J. Mech. Eng. Sci. **205**, 291–304 (1991)
13. Shanyavskiy, A.A., Banov, M.D., Zakharova, T.P.: Principles of physical mesomechanics of nanostructural fatigue of metals. I. Model of subsurface fatigue crack nucleation in VT3-1 titanium alloy. Phys. Mesomech. **13**, 133–142 (2010)
14. Novikov, I.I., Ermishkin, V.A.: On the analysis of stress-strain curves of metals. Izv. Akad. Nauk SSSR. Met. 142–145 (1995)
15. Terent'ev, V.F.: Stage process of fatigue fracture in metallic materials. Izv. Akad. Nauk SSSR. Met. 14–20 (1996)
16. Savkin, A.N.: Estimating the life of material with an irregular load. Russ. Eng. Res. **28**, 17–19 (2008)
17. Brown, M.W., Miller, K.J.: A theory for fatigue failure under multiaxial stress-strain conditions. Proc. Inst. Mech. Eng. **187**, 745–755 (1973)
18. McLean, D.: Mechanical Properties of Metals. Wiley, New York (1962)
19. Griffits, A.A.: The phenomena of rupture and flow in solids. Masinovedenie, 9–14 (1995)
20. Avery, D.H., Backofen, W.A.: Nucleation and growth of fatigue cracks. Fract. Solids **339** (1963)
21. Yokobori, T., Yoshida, M.: Kinetic theory approach to fatigue crack propagation in terms of dislocation dynamics. Int. J. Fract. **10**, 467–470 (1974)
22. Bullen, F.P., Head, A.K., Wood, W.A.: Structural changes during the fatigue of metals. Proc. R. Soc. London. Ser. A. Math. Phys. Sci. **216**, 332–343 (1953)
23. Honeycombe, R.W.K.: The Plastic Deformation of Metals (Subsequent edn.). Hodder Arnold (1984)
24. Karpov, E.V., Demeshkin, A.G., Kornev, V.M.: Damage accumulation in the prefracture zone under non-stationary low-cyclic loading of specimens with the edge crack. 12th Int. Conf. Fract. 2009, ICF-12 **2**, 1195–1204 (2009)
25. Glickman, L.A., Tekht, V.P.: On the question of the physical nature of the progress of metal fatigue. Some Probl. Fatigue Strength Steel, 5–28 (1953)
26. Ronay, M.: Fatigue performance of high strength steels as related to their transformation mechanism. 99–109 (1970)
27. Finkel, V.M.: Physics of Failure. Metallurgiya, Moscow (1970)
28. Panin, V.E.: Methodology of physical mesomechanics as a basis for model construction in computer-aided design of materials. Russ. Phys. J. **38**, 1117–1131 (1995)
29. Peitgen, H.O., Richter, P.: The Beauty of Fractals. Springer-Verlag, Berlin Heidelberg (1986)
30. Panin, V.E., Elsukova, T.F., Angelova, G.V., Kuznetsov, P.V.: Mechanism of formation of fractal mesostructure at the surface of polycrystals upon cyclic loading. Fiz. Met. i Metalloved. **94**, 92–103 (2002)
31. Zhou, X.Y., Chen, D.L., Ke, W., Zang, Q.S., Wang, Z.G.: Fractal characteristics of pitting under cyclic loading. Mater. Lett. **7**, 473–476 (1989)
32. Fellows, J.A.: Fractography and Atlas of Fractographs. Metals Handbook. American Society for Metals, Metals Park, OH (1974)

References

33. Romaniv, O.N., Andrusiv, B.N., Borsukevich, V.I.: Crack formation in fatigue of metals (review). Sov. Mater. Sci. **24**, 1–10 (1988)
34. Laird, C., Smith, G.C.: Crack propagation in high stress fatigue. Philos. Mag. A J. Theor. Exp. Appl. Phys. **7**, 847–857 (1962)
35. Laird, C.: The influence of metallurgical structure on the mechanisms of fatigue crack propagation. In: Grosskreutz, J. (ed.) Fatigue Crack Propagation, pp. 131–180. ASTM International, West Conshohocken, PA, (1967). https://doi.org/10.1520/stp47230s
36. Panin, V.E., Panin, A.V.: Effect of the surface layer in a solid under deformation. Fiz. Mezomekhanika **8**, 7–15 (2005)
37. Bagmutov, V.P., Parshev, S.N.: Integrated approach to the electromechanical formation of a structurally inhomogeneous surface layer on steel parts. Steel Transl. **34**, 66–69 (2004)
38. Bagmutov, V.P., Parshev, S.N., Polozenko, N.Y.: Improvement of mechanical characteristics of a cutting edge of the blade instrument by means of electromechanical processing. Mechanika **56**, 18–20 (2005)
39. Bagmutov, V.P., Parshev, S.N.: Integrated concept of formation of structurally nonuniform surface layer of steel products by electromechanical treatment. Izv. Ferr. Metall. 69–71 (2004)
40. Pokorska, I.: Properties of composite layers obtained by combined treatment. Met. Sci. Heat Treat. **47**, 520–521 (2005)
41. Chudina, O.V., Borovskaya, T.M.: Surface hardening of steels by alloying with laser heating and subsequent chemical heat treatment. Met. Sci. Heat Treat. **39**, 285–287 (1997)
42. Zhukeshov, A.M., Gabdullina, A.T., Amrenova, A.U., Ibraimova, S.A.: Hardening of structural steel by pulsed plasma treatment. J. Nano-Electron. Phys. **6**, (2014)
43. Song, L.X., Zhang, K.M., Zou, J.X., Yan, P.: Surface modifications of a hyperperitectic Zn-10 wt% Cu alloy by pulsed electron beam treatment. Surf. Coatings Technol. **388**, 125530 (2020)
44. Zaguliaev, D., Konovalov, S., Ivanov, Y., Gromov, V., Petrikova, E.: Microstructure and mechanical properties of doped and electron-beam treated surface of hypereutectic Al-11.1% Si alloy. J. Mater. Res. Technol. **8**, 3835–3842 (2019)
45. Ivanov, Y., et al.: Modification of surface layer of hypoeutectic silumin by electroexplosion alloying followed by electron beam processing. Mater. Lett. **253**, 55–58 (2019)
46. Zhang, C., et al.: The microstructure and properties of nanostructured Cr-Al alloying layer fabricated by high-current pulsed electron beam. Vacuum **167**, 263–270 (2019)
47. Zhang, L., Wang, C., Han, L., Dong, C.: Influence of laser power on microstructure and properties of laser clad Co-based amorphous composite coatings. Surfaces and Interfaces **6**, 18–23 (2017)
48. Maruschak, P., Zakiev, I., Mocharsky, V., Nikiforov, Y.: Experimental study of the surface of steel 15Kh13MF after the nanosecond laser shock processing. Solid State Phenom. **200**, 60–65 (2013)
49. Maruschak, P.O., Mocharskyi, V.S., Zakiev, I.M., Nikiforov, Y.M.: Morphology of periodical structures on surface of steel 15Kh13MF after the nanosecond laser irradiation accompanied by generation of shock waves. In: 2012 IEEE International Conference on Oxide Materials for Electronic Engineering (OMEE), pp. 192–193. IEEE (2012). https://doi.org/10.1109/omee.2012.6464902
50. Belyj, A.V., Kukareko, V.A., Sharkeev, Y.P., Panin, S.V., Legostaeva, E.V.: The surface engineering and triboengineering features of 40Kh steel implanted with nitrogen ions. Trenie Iznos **23**, 268–280 (2002)
51. Ovchinnikov, V.V., Borovin Yu, M., Lukyanenko, E.V., Uchevatkina, N.V., Yakutina, S.V.: Study of surface layers obtained by copper ion implantation into a target of steel 30ХГСН2а by auger Spectroscopy methods. Int. J. Eng. Technol. **7**, 93–102 (2018)
52. Fenner, D.B., Hirvonen, J.K., Demaree, J.D.: Selected topics in ion beam surface engineering. In: Engineering Thin Films and Nanostructures with Ion Beams (2005)
53. Zagulyaev, D., et al.: Structure and properties changes of Al-Si alloy treated by pulsed electron beam. Mater. Lett. **229**, 377–380 (2018)
54. Ovchinnikov, V.V., Borovin, J.M., Luk'janenko, E.V., Jakutina, S.V., Uchevatkina, N.V.: Implantation method for surfaces of parts from structural steel. RU 2529337 (2013)

55. Shitsyn, Y.D., Belinin, D.S., Neulybin, S.D., Kuchev, P.S.: Obtaining hardened layers of heat-resistant steels by plasma-welding deposition of congeneric materials. Mod. Appl. Sci. **9**, 64–75 (2015)
56. Yakushin, V.L.: Modification of carbon and low-alloy steels by high-temperature pulsed plasma fluxes. Russ. Metall. **2005**, 104–114 (2005)
57. Pogrebnjak, A., Bazyl, E.: Modification of wear and fatigue characteristics of Ti–V–Al alloy by Cu and Ni ion implantation and high-current electron beam treatment. Vacuum **64**, 1–7 (2001)
58. Chudina, O.V., Petrova, L.G., Borovskaya, T.M.: Mechanisms of hardening of iron by laser alloying and nitriding. Met. Sci. Heat Treat. **44**, 154–159 (2002)
59. Chudina, O.V., Petrova, L.G., Borovskaya, T.M.: The hardening mechanisms of iron upon laser alloying and nitriding. Metalloved. i Termicheskaya Obrab. Met. **22–23** (2002)
60. Kuznetsov, P.V., Oksogoev, A.A., Petrakova, I.V.: The fractal analysis of surface images of aluminum alloy polycrystals subjected to grit blasting in active tension and their fatigue strength. Phys. Mesomech. **7**, 49–57 (2004)
61. Yang, S., et al.: Surface microstructures and high-temperature high-pressure corrosion behavior of N18 zirconium alloy induced by high current pulsed electron beam irradiation. Appl. Surf. Sci. **484**, 453–460 (2019)
62. Hao, S.Z., et al.: WC/Co composite surface structure and nano graphite precipitate induced by high current pulsed electron beam irradiation. Appl. Surf. Sci. **285**, 552–556 (2013)
63. Engelko, V., Yatsenko, B., Mueller, G., Bluhm, H.: Pulsed electron beam facility (GESA) for surface treatment of materials. Vacuum **62**, 211–216 (2001)
64. Han, B., et al.: The phase and microstructure changes in 45# steel irradiated by intense pulsed ion beams. Surf. Coatings Technol. **128–129**, 387–393 (2000)
65. Kormyshev, V.E., Ivanov, Y.F., Gromov, V.E., Konovalov, S.V., Semin, A.P.: Intense pulsed electron beam modification of surface layer facing formed on hardox 450 steel by electrocontact method. J. Surf. Investig. X-ray, Synchrotron Neutron Tech. **11**, 1342–1347 (2017)
66. Kormyshev, V.E., Ivanov, Y.F., Gromov, V.E., Konovalov, S.V., Teresov, A.D.: Surface nanohardness of wear resistant surfacing irradiated by electron beam. Izv. Viss. Uchebnykh Zaved. Chernaya Metall. **60**, 304–309 (2017)
67. Mueller, G., Engelko, V., Weisenburger, A., Heinzel, A.: Surface alloying by pulsed intense electron beams. Vacuum **77**, 469–474 (2005)
68. Ozur, G.E., Proskurovsky, D.I., Rotshtein, V.P., Markov, A.B.: Production and application of low-energy, high-current electron beams. Laser Part. Beams **21**, 157–174 (2003)
69. Rotshtein, V.: Microstructure of the near-surface layers of austenitic stainless steels irradiated with a low-energy, high-current electron beam. Surf. Coatings Technol. **180–181**, 382–386 (2004)
70. Konovalov, S., Ivanov, Y., Gromov, V., Panchenko, I.: Fatigue-induced evolution of AISI 310S steel microstructure after electron beam treatment. Materials **13**(20), 1–13 (2020)
71. Gromov, V.E., Ivanov, Y.F., Vorobiev, S.V., Konovalov, S.V.: Fatigue of steels modified by high intensity electron beams. In: Fatigue of Steels Modified by High Intensity Electron Beams, pp. 1–308. Cambridge International Science Publishing, London (2015)
72. Guan, Q.F., Yang, P.L., Zou, H., Zou, G.T.: Nanocrystalline and amorphous surface structure of 0.45%C steel produced by high current pulsed electron beam. J. Mater. Sci. **41**, 479–483 (2006)
73. Hao, S.Z., et al.: Fundamentals and applications of material modification by intense pulsed beams. Surf. Coatings Technol. **201**, 8588–8595 (2007)
74. Hao, S., et al.: Surface modification of steels and magnesium alloy by high current pulsed electron beam. Nucl. Instruments Methods Phys. Res. Sect. B Beam Interact. with Mater. Atoms **240**, 646–652 (2005)
75. Hu, J.J., Zhang, G.B., Xu, H.B., Chen, Y.F.: Microstructure characteristics and properties of 40Cr steel treated by high current pulsed electron beam. Mater. Technol. (2012). https://doi.org/10.1179/175355511X13171168481358

References

76. Dong, C., et al.: Surface treatment by high current pulsed electron beam. Surf. Coatings Technol. **163–164**, 620–624 (2003)
77. Qin, Y., et al.: Temperature–stress fields and related phenomena induced by a high current pulsed electron beam. Nucl. Instruments Methods Phys. Res. Sect. B Beam Interact. with Mater. Atoms **225**, 544–554 (2004)
78. Grosdidier, T., Zou, J.X., Bolle, B., Hao, S.Z., Dong, C.: Grain refinement, hardening and metastable phase formation by high current pulsed electron beam (HCPEB) treatment under heating and melting modes. J. Alloys Compd. **504**, S508–S511 (2010)
79. Guan, Q.F., et al.: Surface nanostructure and amorphous state of a low carbon steel induced by high-current pulsed electron beam. Surf. Coatings Technol. **196**, 145–149 (2005)
80. Zou, J., et al.: Oxidation protection of AISI H13 steel by high current pulsed electron beam treatment. Surf. Coatings Technol. **183**, 261–267 (2004)
81. Zhang, K.M., Zou, J.X., Grosdidier, T., Dong, C., Weber, S.: Rapid surface alloying by Ti of AISI 316L stainless steel using. Low energy high current pulsed electron beam. Eur. Phys. J. Appl. Phys. **43**, 343–347 (2008)
82. Zhang, K., Zou, J., Grosdidier, T., Dong, C., Yang, D.: Improved pitting corrosion resistance of AISI 316L stainless steel treated by high current pulsed electron beam. Surf. Coatings Technol. **201**, 1393–1400 (2006)
83. Zou, J.X., Grosdidier, T., Zhang, K.M., Dong, C.: Cross-sectional analysis of the graded microstructure in an AISI D2-steel treated with low energy high-current pulsed electron beam. Appl. Surf. Sci. **255**, 4758–4764 (2009)
84. Zou, J.X., et al.: Orientation-dependent deformation on 316L stainless steel induced by high-current pulsed electron beam irradiation. Mater. Sci. Eng., A **483–484**, 302–305 (2008)
85. Zou, J.X., Zhang, K.M., Hao, S.Z., Dong, C., Grosdidier, T.: Mechanisms of hardening, wear and corrosion improvement of 316L stainless steel by low energy high current pulsed electron beam surface treatment. Thin Solid Films **519**, 1404–1415 (2010)
86. Hao, S., Wu, P., Zou, J., Grosdidier, T., Dong, C.: Microstructure evolution occurring in the modified surface of 316L stainless steel under high current pulsed electron beam treatment. Appl. Surf. Sci. **253**, 5349–5354 (2007)
87. Xu, F., Tang, G., Guo, G., Ma, X., Ozur, G.E.: Influence of irradiation number of high current pulsed electron beam on the structure and properties of M50 steel. Nucl. Instruments Methods Phys. Res. Sect. B Beam Interact. with Mater. Atoms **268**, 2395–2399 (2010)
88. Tang, G.Z., Xu, F.J., Ma, X.X.: Microstructure and properties of surface modified layer for M50 steel by high current pulsed electron beam. Jinshu Rechuli/Heat Treat. Met. (2010)
89. Zou, J.X., et al.: Microstructures and phase formations in the surface layer of an AISI D2 steel treated with pulsed electron beam. J. Alloys Compd. (2007). https://doi.org/10.1016/j.jallcom.2006.08.280
90. Zhang, K., Zou, J., Grosdidier, T., Dong, C.: Crater-formation-induced metastable structure in an AISI D2 steel treated with a pulsed electron beam. Vacuum **86**, 1273–1277 (2012)
91. Zou, J., Grosdidier, T., Zhang, K., Dong, C.: Mechanisms of nanostructure and metastable phase formations in the surface melted layers of a HCPEB-treated D2 steel. Acta Mater. **54**, 5409–5419 (2006)
92. Zhang, K.M., Zou, J.X., Bolle, B., Grosdidier, T.: Evolution of residual stress states in surface layers of an AISI D2 steel treated by low energy high current pulsed electron beam. Vacuum **87**, 60–68 (2013)
93. Gao, Y.: Influence of pulsed electron beam treatment on microstructure and properties of TA15 titanium alloy. Appl. Surf. Sci. **264**, 633–635 (2013)
94. Gao, Y.: Surface modification of TC4 titanium alloy by high current pulsed electron beam (HCPEB) with different pulsed energy densities. J. Alloys Compd. **572**, 180–185 (2013)
95. Gao, Y.K.: Surface modification of TA2 pure titanium by low energy high current pulsed electron beam treatments. Appl. Surf. Sci. (2011). https://doi.org/10.1016/j.apsusc.2011.03.005
96. Guo, G., Tang, G., Ma, X., Sun, M., Ozur, G.E.: Effect of high current pulsed electron beam irradiation on wear and corrosion resistance of Ti6Al4 V. Surf. Coatings Technol. (2013). https://doi.org/10.1016/j.surfcoat.2012.08.009

97. Zhang, X.D., et al.: Microstructure and property modifications in a near α Ti alloy induced by pulsed electron beam surface treatment. Surf. Coatings Technol. **206**, 295–304 (2011)
98. Zhang, K.M., Yang, D.Z., Zou, J.X., Grosdidier, T., Dong, C.: Improved in vitro corrosion resistance of a NiTi alloy by high current pulsed electron beam treatment. Surf. Coatings Technol. **201**, 3096–3102 (2006)
99. Zhang, K. M., et al.: Surface modification of Ni (50.6at.%) Ti by high current pulsed electron beam treatment. J. Alloys Compd. **434–435**, 682–685 (2007)
100. Gao, B., et al.: Effect of high current pulsed electron beam treatment on surface microstructure and wear and corrosion resistance of an AZ91HP magnesium alloy. Surf. Coatings Technol. **201**, 6297–6303 (2007)
101. Gao, B., et al.: High current pulsed electron beam treatment of AZ31 Mg alloy. J. Vac. Sci. Technol. A Vacuum, Surfaces, Film **23**, 1548–1553 (2005)
102. Gao, B., Hao, S.Z., Dong, C., Tu, G.F.: Improvement of wear resistance of AZ31 and AZ91HP by high current pulsed electron beam treatment. Trans. Nonferrous Met. Soc. China (English Ed.) **15**, 978–984 (2005)
103. Gao, B., et al.: Improvement of wear resistance of magnesium alloy AZ91HP by high current pulsed electron beam treatment. Transactions Mater. Heat Treat. **25**, 1029–1031 (2004)
104. Hao, S., He, D., Zhao, L.: Microstructure and corrosion resistance of FeCrAl coating after high current pulsed electron beam surface modification. Procedia Eng. **27**, 1700–1706 (2012)
105. Hao, S., Zhao, L., He, D.: Surface microstructure and high temperature corrosion resistance of arc-sprayed FeCrAl coating irradiated by high current pulsed electron beam. Nucl. Instruments Methods Phys. Res. Sect. B Beam Interact. with Mater. Atoms **312**, 97–103 (2013)
106. Hao, S., Zhang, X., Mei, X., Grosdidier, T., Dong, C.: Surface treatment of DZ4 directionally solidified nickel-based superalloy by high current pulsed electron beam. Mater. Lett. **62**, 414–417 (2008)
107. Hao, S., Xu, Y., Zhang, Y., Zhao, L.: Improvement of surface microhardness and wear resistance of WC/Co hard alloy by high current pulsed electron beam irradiation. Int. J. Refract. Met. Hard Mater. **41**, 553–557 (2013)
108. Xu, Y., et al.: Surface microstructure and mechanical property of WC-6% Co hard alloy irradiated by high current pulsed electron beam. Appl. Surf. Sci. (2013). https://doi.org/10.1016/j.apsusc.2013.04.053
109. Uglov, V.V., et al.: Structure, phase composition and mechanical properties of hard alloy treated by intense pulsed electron beams. Surf. Coatings Technol. **206**, 2972–2976 (2012)
110. Koval, N.N., et al.: Structure and properties of the surface alloy formed by irradiating a film/substrate system with a high-intensity electron beam. Russ. Phys. J. **54**, 1024–1033 (2012)
111. Uglov, V.V., et al.: Structure-phase transformation in surface layers of hard alloy as a result of action of high-current electron beams. J. Surf. Investig. X-ray, Synchrotron Neutron Tech. **5**, 350–357 (2011)
112. Koval, N.N., et al.: Surface modification of TiC-NiCrAl hard alloy by pulsed electron beam. IEEE Trans. Plasma Sci. **37**, 1998–2001 (2009)
113. Kolubaeva, Y.A., Ivanov, Y.F., Devyatkov, V.N., Koval, N.N.: Pulsed-periodic electron-beam treatment of quenched steel. Steel Transl. **37**, 662–665 (2007)
114. Fedun, V.I., Kolyada, Y.E.: Dynamics of phase transformations by electron beem surface modification of metals and alloys. Probl. At. Sci. Technol. 316–320 (2010)
115. Boyer, R.R.: Aerospace applications of beta titanium alloys. JOM **46**, 20–23 (1994)
116. Gromov, V.E., et al.: Increase in fatigue life of steels by electron-beam processing. J. Surf. Investig. X-ray, Synchrotron Neutron Tech. **10**, 83–87 (2016)
117. Gromov, V.E., Ivanov, Y.F., Sizov, V.V., Vorob'ev, S.V., Konovalov, S.V.: Increase in the fatigue durability of stainless steel by electron-beam surface treatment. J. Surf. Investig. X-ray, Synchrotron Neutron Tech. **7**, 94–98 (2013)
118. Ivanov, Y.F., et al.: On the fatigue strength of grade 20Cr13 hardened steel modified by an electron beam. J. Surf. Investig. X-ray, Synchrotron Neutron Tech. **7**, 90–93 (2013)

119. Sizov, V.V., Gromov, V.E., Ivanov, Y.F., Vorob'ev, S.V., Konovalov, S.V.: Fatigue failure of stainless steel after electron-beam treatment. Steel Transl. **42**, 486–488 (2012)
120. Vorob'ev, S.V., Gromov, V.E., Ivanov, Y.F., Sizov, V.V., Sofroshenkov, A.F.: Nanocrystalline structure and fatigue life of stainless steel. Steel Transl. **42**, 316–318 (2012)
121. Gromov, V.E., Gorbunov, S.V., Ivanov, Y.F., Vorobiev, S.V., Konovalov, S.V.: Formation of surface gradient structural-phase states under electron-beam treatment of stainless steel. J. Surf. Investig. X-ray, Synchrotron Neutron Tech. **5**, 974–978 (2011)
122. Ivanov, Y.F., et al.: Multicyclic fatigue of stainless steel treated by a high-intensity electron beam: surface layer structure. Russ. Phys. J. **54**, 575–583 (2011)
123. Ivanov, Y.F., Gromov, V.E., Gorbunov, S.V., Vorob'ev, S.V., Konovalov, S.V.: Gradient structural phase states formed in steel 08Kh18N10T in the course of high-cycle fatigue to failure. Phys. Met. Metallogr. **112**, 81–89 (2011)
124. Grishunin, V.A., Gromov, V.E., Ivanov, Y.F., Volkov, K.V., Konovalov, S.V.: Evolution of the phase composition and defect substructure in the surface layer of rail steel under fatigue. Steel Transl. **43**, 724–727 (2013)
125. Grishunin, V.A., Gromov, V.E., Ivanov, Y.F., Teresov, A.D., Konovalov, S.V.: Evolution of the phase composition and defect substructure of rail steel subjected to high-intensity electron-beam treatment. J. Surf. Investig. X-ray, Synchrotron Neutron Tech. **7**, 990–995 (2013)
126. Gromov, V.E., Konovalov, S.V., Aksenova, K.V., Kobzareva, T.Y.: The Evolution of the Structure and Properties of Light Alloys at the Energy Influences. Publishing house of the Siberian Branch of the Russian Academy of Sciences, Novosibirsk (2016)
127. Gromova, A.V., et al.: Ways of the dislocation substructure evolution in austenite steel under low and multicycle fatigue. Procedia Eng. (2010). https://doi.org/10.1016/j.proeng.2010.03.009
128. Sharkeev, Y.P., Kukareko, V.A., Legostaeva, E.V., Byeli, A.V.: Surface-hardened nanostructured Ti- and Zr-matrix composites for medical and engineering applications. Russ. Phys. J. **53**, 1053–1059 (2011)
129. Ghosh, S.K.: Functional Coatings: By Polymer Microencapsulation. Wiley, New York (2006). https://doi.org/10.1002/3527608478
130. Pietrowski, S.: Characteristic features of silumin alloys crystallization. Mater. Des. **18**, 379–383 (1997)
131. Zolotorevsky, V.S., Belov, N.A., Glazoff, M.V.: Casting Aluminum Alloys. Elsevier, Amsterdam (2007). https://doi.org/10.1016/B978-0-08-045370-5.X5001-9
132. Belov, N.A., Alabin, A.N.: Use of multicomponent phase diagrams for designing high strength casting aluminum alloys. Mater. Sci. Forum (2014). https://doi.org/10.4028/www.scientific.net/MSF.794-796.909
133. Muratov, V.S., Morozova, E.A.: Titanium structure and property formation with chromium laser surface alloying. Met. Sci. Heat Treat. **61**, 340–343 (2019)
134. Glavatskikh, M.V., Pozdniakov, A.V., Makhov, S.V., Napalkov, V.I.: Investigation into the structure and phase composition of powder aluminum-phosphorus master alloys. Russ. J. Non-Ferrous Met. **55**, 450–455 (2014)
135. Pozdnyakov, A.V., Osipenkova, A.A., Popov, D.A., Makhov, S.V., Napalkov, V.I.: Effect of low additions of Y, Sm, Gd, Hf and Er on the structure and hardness of alloy Al–0.2% Zr–0.1% Sc. Met. Sci. Heat Treat. **58**, 537–542 (2017)
136. Elmadagli, M., Perry, T., Alpas, A.T.: A parametric study of the relationship between microstructure and wear resistance of Al–Si alloys. Wear **262**, 79–92 (2007)
137. Zeren, M., Karakulak, E.: Influence of Ti addition on the microstructure and hardness properties of near-eutectic Al–Si alloys. J. Alloys Compd. **450**, 255–259 (2008)
138. Robles Hernádez, F.C., Sokolowski, J.H.: Thermal analysis and microscopical characterization of Al-Si hypereutectic alloys. J. Alloys Compd. **419**, 180–190 (2006)
139. Zeren, M.: The effect of heat-treatment on aluminum-based piston alloys. Mater. Des. **28**, 2511–2517 (2007)
140. Taghiabadi, R., Ghasemi, H.M., Shabestari, S.G.: Effect of iron-rich intermetallics on the sliding wear behavior of Al–Si alloys. Mater. Sci. Eng., A (2008). https://doi.org/10.1016/j.msea.2008.01.001

141. Li, R., et al.: Age-hardening behavior of cast Al–Si base alloy. Mater. Lett. **58**, 2096–2101 (2004)
142. Belov, N.A., Eskin, D.G., Aksenov, A.A.: Multicomponent Phase Diagrams. Multicomponent Phase Diagrams: Applications for Commercial Aluminum Alloys. Elsevier, Amsterdam (2005). https://doi.org/10.1016/b978-0-08-044537-3.x5000-8
143. Mohamed, A.M.A., Samuel, A.M., Samuel, F.H., Doty, H.W.: Influence of additives on the microstructure and tensile properties of near-eutectic Al–10.8%Si cast alloy. Mater. Des. **30**, 3943–3957 (2009)
144. Pogrebnyak, A.D., Tyurin, Y.N.: Modification of material properties and coating deposition using plasma jets. Phys. Usp. **48**, 487–514 (2005)
145. Tsvirkun, O.A., Bagautdinov, A.Y., Ivanov, Y.F., Budovskikh, E.A., Gromov, V.E.: Phase composition and defect substructure of nickel alloyed with boron and copper by electric explosion of conductors. Russ. Phys. J. **50**, 199–203 (2007)
146. Girzhon, V.V., Kovalyova, V.M., Smolyakov, O.V.: Decagonal quasi-crystalline phase formation during laser alloying of aluminium with cobalt and nickel. Metallofiz. I Noveishie Tekhnologii. **36**, 745–756 (2016)
147. Girzhon, V.V., Kovalyova, V.M., Smolyakov, O.V.: Surface-layers' structure of hypoeutectic silumin after laser alloying by mixture of the copper and cobalt powders. Metallofiz. I Noveishie Tekhnologii. **37**, 703–712 (2016)
148. Wong, T.T., Liang, G.Y.: Effect of laser melting treatment on the structure and corrosion behaviour of aluminium and AlSi alloys. J. Mater. Process. Technol. **63**, 930–934 (1997)
149. Sicard, E., Boulmer-Leborgne, C., Andreazza-Vignolle, C., Frainais, M.: Excimer laser surface treatment of aluminum alloy in nitrogen. Appl. Phys. A Mater. Sci. Process. **73**, 55–60 (2001)
150. Chong, P.H., Liu, Z., Skeldon, P., Thompson, G.E.: Large area laser surface treatment of aluminium alloys for pitting corrosion protection. Appl. Surf. Sci. (2003). https://doi.org/10.1016/S0169-4332(02)01418-6
151. Tomida, S., Nakata, K., Shibata, S., Zenkouji, I., Saji, S.: Improvement in wear resistance of hyper-eutectic Al-Si cast alloy by laser surface remelting. Surf. Coatings Technol. **169–170**, 468–471 (2003)
152. Brekharya, G.P., Girzhon, V.V., Tantsyura, I.V.: Formation of structure of surface layers of the eutectic and hypereutectic silumins after pulse laser processing. Metallofiz. i Noveishie Tekhnologii. **27**, 1519–1525 (2007)
153. Wong, T.T., Liang, G.Y., Tang, C.Y.: The surface character and substructure of aluminium alloys by laser-melting treatment. J. Mater. Process. Technol. **66**, 172–178 (1997)
154. Akamatsu, H., et al.: Increase of Si solution rate into Al matrix by repeated irradiation of intense pulsed ion beam. Vacuum **65**, 563–569 (2002)
155. Rygina, M.E., et al.: Modification of hypereutectic silumin by ion-electron-plasma method. Key Eng. Mater. **769**, 54–59 (2018)
156. Ivanov, Y.F., et al.: Modification of the silumin structure and properties by electron-ionplasma saturation of the surface with atoms of metals and gases. High Temp. Mater. Process. An Int. Q. High-Technology Plasma Process. **20**, 295–307 (2016)
157. Ivanov, Y.F., et al.: Modification of hypereutectic silumin surface layer by a high-intensity pulse electron beam. High Temp. Mater. Process. (An Int. Q. High-Technology Plasma Process.) **19**, 85–91 (2015)
158. Hao, Y., et al.: Surface modification of Al–12.6Si alloy by high current pulsed electron beam. Appl. Surf. Sci. **258**, 2052–2056 (2012)
159. Gao, B., et al.: Study on continuous solid solution of Al and Si elements of a high current pulsed electron beam treated hypereutectic Al-17.5Si alloy. Physics Procedia (2011). https://doi.org/10.1016/j.phpro.2011.06.079
160. Gao, B., et al.: High current pulsed electron beam treatment of hypereutectic Al-17.5Si alloy. Cailiao Rechuli Xuebao/Transactions Mater. Heat Treat. **31**, 115–118 (2010)
161. Hao, Y., et al.: Improved wear resistance of Al–15Si alloy with a high current pulsed electron beam treatment. Nucl. Instruments Methods Phys. Res. Sect. B Beam Interact. with Mater. Atoms **269**, 1499–1505 (2011)

162. Hao, Y., et al.: Surface modification of Al–20Si alloy by high current pulsed electron beam. Appl. Surf. Sci. **257**, 3913–3919 (2011)
163. Hao, Y., et al. Effect of HCPEB treatment on microstructure and microhardness of hypereutectic Al-20Si alloy. Cailiao Rechuli Xuebao/Transactions Mater. Heat Treat. 142–145 (2010)
164. Hao, Y., Gao, B., Tu, G.F., Wang, Z., Hao, S.Z.: Influence of high current pulsed electron beam (HCPEB) treatment on wear resistance of hypereutectic Al-17.5Si and Al-20Si alloys. Mater. Sci. Forum. **675–677**, 693–696 (2011)
165. An, J., et al.: Influence of high current pulsed electron beam treatment on the tribological properties of Al–Si–Pb alloy. Surf. Coatings Technol. **200**, 5590–5597 (2006)
166. An, J., Shen, X.X., Lu, Y., Liu, Y.B.: Microstructure and tribological properties of Al–Pb alloy modified by high current pulsed electron beam. Wear **261**, 208–215 (2006)
167. Hao, S., et al.: Surface treatment of aluminum by high current pulsed electron beam. Curr. Appl. Phys. **1**, 203–208 (2001)
168. Grosdidier, T., et al.: Texture modification, grain refinement and improved hardness/corrosion balance of a FeAl alloy by pulsed electron beam surface treatment in the "heating mode". Scr. Mater. **58**, 1058–1061 (2008)
169. Peters, M., Leyens, C.: Titanium and Titanium Alloys. Titanium and Titanium Alloys Fundamentals and Applications. Wiley, New York (2003). https://doi.org/10.1002/3527602119
170. Polmear, I., StJohn, D., Nie, J.F., Qian, M.: Titanium alloys. In: Light Alloys, pp. 369–460. Elsevier (2017). https://doi.org/10.1016/b978-0-08-099431-4.00007-5
171. Yang, X., Richard Liu, C.: Machining titanium and its alloys. Mach. Sci. Technol. **3**, 107–139 (1999)
172. Boyer, R.R., Briggs, R.D.: The use of β titanium alloys in the aerospace industry. J. Mater. Eng. Perform. **22**, 2916–2920 (2013)
173. Titanium and Titanium Alloys. Titanium and Titanium Alloys. Wiley, New York (2003). https://doi.org/10.1002/3527602119
174. Banerjee, D., Williams, J.C.: Perspectives on titanium science and technology. Acta Mater. **61**, 844–879 (2013)
175. Suárez, A., et al.: Modeling of phase transformations of Ti6Al4 V during laser metal deposition. Phys. Procedia **12**, 666–673 (2011)
176. Price, T.S., Shipway, P.H., McCartney, D.G.: Effect of cold spray deposition of a titanium coating on fatigue behavior of a titanium alloy. J. Therm. Spray Technol. **15**, 507–512 (2006)
177. Wong, W., Irissou, E., Ryabinin, A.N., Legoux, J.-G., Yue, S.: Influence of helium and nitrogen gases on the properties of cold gas dynamic sprayed pure titanium coatings. J. Therm. Spray Technol. **20**, 213–226 (2011)
178. Wong, W., et al.: Effect of particle morphology and size distribution on cold-sprayed pure titanium coatings. J. Therm. Spray Technol. **22**, 1140–1153 (2013)
179. Sharma, B., Vajpai, S., Ameyama, K.: An efficient powder metallurgy processing route to prepare high-performance β-Ti–Nb alloys using pure titanium and titanium hydride powders. Metals (Basel) **8**, 516 (2018)
180. Weng, F., Chen, C., Yu, H.: Research status of laser cladding on titanium and its alloys: a review. Mater. Des. **58**, 412–425 (2014)
181. Sampedro, J., Pérez, I., Carcel, B., Ramos, J.A., Amigó, V.: Laser cladding of TiC for better titanium components. Phys. Procedia **12**, 313–322 (2011)
182. Chakraborty, R., Raza, M.S., Datta, S., Saha, P.: Synthesis and characterization of nickel free titanium–hydroxyapatite composite coating over Nitinol surface through in-situ laser cladding and alloying. Surf. Coatings Technol. **358**, 539–550 (2019)
183. Popova, A.A., Yakovlev, V.I., Legostaeva, E.V., Sitnikov, A.A., Sharkeev, Y.P.: The effect of the granulometric composition of a hydroxyapatite powder on the structure and phase composition of coatings deposited by the detonation gas spraying technique. Russ. Phys. J. **55**, 1284–1289 (2013)

184. Koshuro, V.A., Nechaev, G.G., Lyasnikova, A.V.: Composition and structure of coatings formed on a VT16 titanium alloy by electro-plasma spraying combined with microarc oxidation. Tech. Phys. **59**, 1570–1572 (2014)
185. Koshuro, V., Fomin, A., Rodionov, I.: Composition, structure and mechanical properties of metal oxide coatings produced on titanium using plasma spraying and modified by micro-arc oxidation. Ceram. Int. **44**, 12593–12599 (2018)
186. Nechaev, G.G., Popova, S.S.: Dynamic model of single discharge during microarc oxidation. Theor. Found. Chem. Eng. **49**, 447–452 (2015)
187. Koshuro, V.A., Nechaev, G.G., Lyasnikova, A.V.: Effect of plasma processes of coating formation on the structure and mechanical properties of titanium. Inorg. Mater. Appl. Res. **7**, 350–353 (2016)
188. Tian, Y.S., Chen, C.Z., Li, S.T., Huo, Q.H.: Research progress on laser surface modification of titanium alloys. Appl. Surf. Sci. **242**, 177–184 (2005)
189. Luo, G.X., Wu, G.Q., Huang, Z., Ruan, Z.J.: Microstructure transformations of lasersurface-melted near-alpha titanium alloy. Mater. Charact. **60**, 525–529 (2009)
190. Poulon-Quintin, A., Watanabe, I., Watanabe, E., Bertrand, C.: Microstructure and mechanical properties of surface treated cast titanium with Nd:YAG laser. Dent. Mater. **28**, 945–951 (2012)
191. Sun, R., Yang, D., Guo, L., Dong, S.: Microstructure and wear resistance of NiCrBSi laser clad layer on titanium alloy substrate. Surf. Coatings Technol. **132**, 251–255 (2000)
192. Liu, Z.D., Zhang, X.C., Xuan, F.Z., Wang, Z.D., Tu, S.T.: In situ synthesis of TiN/Ti3Al intermetallic matrix composite coatings on Ti6Al4 V alloy. Mater. Des. **37**, 268–273 (2012)
193. Liu, X.-B., et al.: Development and characterization of laser clad high temperature selflubricating wear resistant composite coatings on Ti–6Al–4 V alloy. Mater. Des. **55**, 404–409 (2014)
194. Romanov, D.A., et al.: Structure and phase composition of wear-resistant coatings of the TiB2–Al system prepared by electroexplosion sputtering. Russ. J. Non-Ferrous Met. **57**, 75–79 (2016)
195. Ion, J.C.: Laser Processing of Engineering Materials: Principles, Procedure and Industrial Application. (2005). ISBN:0750660791
196. de Oliveira, U., Ocelík, V., De Hosson, J.T.M.: Analysis of coaxial laser cladding processing conditions. Surf. Coatings Technol. **197**, 127–136 (2005)
197. Moures, F., et al.: Optimisation of refractory coatings realised with cored wire addition using a high-power diode laser. Surf. Coatings Technol. **200**, 2283–2292 (2005)
198. Sarma, B., Ravi Chandran, K.S.: Recent advances in surface hardening of titanium. JOM **63**, 85–92 (2011)
199. Gordienko, P.S., Vasilenko, O.S., Dostovalov, V.A., Dostovalov, D.V.: Formation of protective coatings on titanium alloys. Tsvetnye Met. 65–69 (2017) https://doi.org/10.17580/tsm.2017.01.11
200. Gordienko, P.S., Zhevtun, I.G., Dostovalov, V.A., Kuryavyi, V.G., Barinov, N.N.: Composition and structure of carbon-rich local sections formed on titanium alloys in electrolytes. Russ. Eng. Res. **32**, 158–161 (2012)
201. Wu, G., et al.: Effect of laser surface melting pretreatment on the growth behavior and mechanical properties of microarc oxidation coating on Ti6Al4V alloy. J. Laser Appl. **32**, 012013 (2020)
202. Ivanov, Y.F., et al.: Formation of nanocomposite layers at the surface of VT1-0 titanium in electroexplosive carburization and electron-beam treatment. Steel Transl. **42**, 499–501 (2012)
203. Romanov, D.A., Budovskikh, E.A., Zhmakin, Y.D., Gromov, V.E.: Surface modification by the EVU 60/10 electroexplosive system. Steel Transl. **41**, 464–468 (2011)
204. Budovskikh, E.A., Gromov, V.E., Romanov, D.A.: The formation mechanism providing high-adhesion properties of an electric-explosive coating on a metal basis. Dokl. Phys. **58**, 82–84 (2013)
205. Ivanov, Y.F., et al.: Steel 45 surface modification by a combined electron-ion-plasma method. High Temp. Mater. Process. (An Int. Q. High-Technology Plasma Process). **19**, 29–36 (2015)

References 51

206. Sosnin, K.V., et al.: Combined electron-ion-plasma doping of a titanium surface with yttrium: analyzing structure and properties. Bull. Russ. Acad. Sci. Phys. **78**, 1183–1187 (2014)
207. Soskova, N.A., Budovskikh, E.A., Gromov, V.E., Ivanov, Y.F., Raikov, S.V.: Formation of dislocation-free nanostructures in metals on electroexplosive alloying. Steel Transl. **42**, 820–822 (2012)
208. Poletika, I.M., et al.: Development of a new class of coatings by double electron-beam surfacing. Inorg. Mater. Appl. Res. **2**, 531–539 (2011)
209. Muratov, V.S., Morozova, E.A.: Formation of structure and properties of titanium under laser surface alloying with nickel and manganese. Met. Sci. Heat Treat. **60**, 589–593 (2019)
210. Shetulov, D.I., Kravchenko, V.N., Myl'nikov, V.V.: Predicting the strength and life of auto parts on the basis of fatigue strength. Russ. Eng. Res. **35**, 580–583 (2015)
211. Myl'nikov, V.V., Shetulov, D.I., Chernyshov, E.A.: Investigation into the surface damage of pure metals allowing for the cyclic loading frequency. Russ. J. Non-Ferrous Met. **54**, 229–233 (2013)
212. Agzamov, R.D., Tagirov, A.F., Ramazanov, K.N.: Influence of ion nitriding regimes on diffusion processes in titanium alloy Ti-6Al-4V. Defect Diffus. Forum. **383**, 161–166 (2018)
213. Ivanov, Y.F., et al.: Mechanisms of low-temperature diffusion saturation of structural steel with nitrogen in the plasma of a low-pressure arc discharge. High Temp. Mater. Process. (An Int. Q. High-Technology Plasma Process). **17**, 213–219 (2013)
214. Khusainov, Y.G., Ramazanov, K.N.: Local ion nitriding of martensitic structural steel in plasma of glow discharge with hollow cathode. Inorg. Mater. Appl. Res. **10**, 544–548 (2019)
215. Krivobokov, V., Stepanova, O., Yuryeva, A.: Erosion yield of metal surface under ion pulsed irradiation. Nucl. Instruments Methods Phys. Res. Sect. B Beam Interact. with Mater. Atoms **315**, 261–264 (2013)
216. Bagautdinov, A.Y., Budovskikh, E.A., Ivanov, Y.F., Martusevich, E.V., Gromov, V.E.: Mesostructural level of nickel modification by boron during surface treatment by electrical explosion. Fiz. Mezomekhanika **8**, 89–94 (2005)
217. Cherenda, N.N., et al.: Modification of Ti-6Al-4V alloy element and phase composition by compression plasma flows impact. Surf. Coatings Technol. **355**, 148–154 (2018)
218. Golkovski, M.G., et al.: Atmospheric electron-beam surface alloying of titanium with tantalum. Mater. Sci. Eng., A **578**, 310–317 (2013)
219. Yun, E., Lee, K., Lee, S.: Correlation of microstructure with high-temperature hardness of (TiC, TiN)/Ti–6Al–4V surface composites fabricated by high-energy electron-beam irradiation. Surf. Coatings Technol. **191**, 83–89 (2005)
220. Ivanov, Y.F., Kobzareva, T.A., Gromov, V.E., Budovskikh, E.A., Bashchenko, L.P.: Electroexplosive doping of titanium alloy by boron carbide and subsequent electron beam processing. High Temp. Mater. Process. (An Int. Q. High-Technology Plasma Process). **18**, 281–290 (2014)
221. Yun, E., Lee, K., Lee, S.: Improvement of high-temperature hardness of (TiC, TiB)/Ti–6Al–4V surface composites fabricated by high-energy electron-beam irradiation. Surf. Coatings Technol. **184**, 74–83 (2004)
222. Xue, Y., Wang, H.M.: Microstructure and wear properties of laser clad TiCo/Ti2Co intermetallic coatings on titanium alloy. Appl. Surf. Sci. **243**, 278–286 (2005)
223. Bataeva, E.A., Bataev, I.A., Burov, V.G., Tushinskii, L.I., Golkovskii, M.G.: Effect of initial state on inhomogeneity of the structure of carbon steels hardened by electron-beam treatment at atmospheric pressure. Met. Sci. Heat Treat. **51**, 103–105 (2009)
224. Demina, E.V., et al.: Changes in the structure and nitride phase formation in the surface layer of Fe-based alloys under the pulsed action of nitrogen ions and nitrogen plasma. Inorg. Mater. Appl. Res. **2**, 230–236 (2011)
225. Ivanov, L.I., et al.: Diffusion loss of hydrogen isotopes from iron based alloys. Defect Diffus. Forum. **194–199**, 1093–1098 (2001)
226. Ugaste, Ü.: The concentration dependence of diffusion coefficients in binary metal systems: empirical relationships. Defect Diffus. Forum. **194–199**, 157–162 (2001)

227. Sizov, I., Mishigdorzhiyn, U., Leyens, C., Vetter, B., Fuhrmann, T.: Influence of thermocycle boroaluminising on strength of steel C30. Surf. Eng. **30**, 129–133 (2014)
228. Rotshtein, V., Ivanov, Y., Markov, A.: Surface treatment of materials with low-energy, high-current electron beams. In: Materials Surface Processing by Directed Energy Techniques, pp. 205–240. Elsevier, Amsterdam (2006). https://doi.org/10.1016/b978-008044496-3/50007-1
229. Kraposhin, V.S.: Engineering relations for the depth of surface heating of metal by highly concentrated energy sources. Met. Sci. Heat Treat. **41**, 301–305 (1999)
230. Vostretsova, A.V., Karpii, S.V., Budovskikh, E.A., Goryushkin, V.F., Gromov, V.E.: Carbidization of titanium alloys in electroexplosive carburization and additional heat treatment. Steel Transl. **39**, 466–469 (2009)
231. Karpii, S.V., Morozov, M.M., Ivanov, Y.F., Budovskikh, E.A., Gromov, V.E.: Formation of nanophases in electroexplosive alloying with aluminum and boron and electronbeam treatment of titanium surfaces. Steel Transl. **40**, 723–728 (2010)
232. Ovcharenko, V.E., Psakh'e, S.G., Boyangin, E.N.: Formation of a multimodal grain structure and its influence on the strength and the ductility of the intermetallic compound Ni3Al. Russ. Metall. **2014**, 299–302 (2014)
233. Mikheev, R.S., Chernyshova, T.A.: Processing of the surface of a silumin plate with a high-energy source. Russ. Metall. **2009**, 498–504 (2009)
234. Khaimzon, B.B., Sarychev, V.D., Soskova, N.A., Gromov, V.E.: Temperature distribution produced by pulsed energy fluxes, with evaporation of the target. Steel Transl. **43**, 55–58 (2013)

Chapter 2
Special Analysis Aspects of Modified Light Alloys

2.1 Materials of Research

To analyze the fatigue, researchers have tested commercial as-cast alloy of aluminum and silicon AK12 (eutectic silumin) [1, 2] and alloy VT1-0, the chemical compositions of which are given in Table 2.1.

Titanium-based alloy VT6 was utilized as a hardener in the experimenting on face hardening of materials. Table 2.2 provides the data of chemical composition of the alloy meeting the specifications of GOST (Russian State Standard) 19807—91. Samples were prepared in a form of disks (5 mm thick, 18 mm in diameter).

2.2 Methods of Fatigue Tests

Fatigue tests, similar as described before [3–11], were carried out using a specially assembled experimental unit and application of asymmetrical cyclic cantilever bending. Specimens were prepared in a form of a parallelepiped (8 mm × 14 mm × 145 mm) for silumin and 12 mm × 4 mm × 130 mm for titanium (Fig. 2.1). To imitate a crack, a semi-sphere was cut with radii of 10 mm and 20 mm, respectively. The highest possible stress (10 MPa) of a loading cycle was found experimentally, so as a sample could withstand a load within at least 10^5 cycles until fracturing. All tests were conducted at room temperature (≈ 300 K). The frequency of bending loads was selected to be 15 Hz for silumin and 20 Hz for titanium. The purpose of the tests was to find out a possible number of cycles till the fracture of samples. Figure 2.2 shows an external view of an experimental unit for fatigue tests.

A sample (1) is clamped (2). One tip of the sample is fixed, whereas an alternating load is applied to the other one. The lever mechanism (3) connected to the shaft (8) initiates the bending. The eccentric (6) is employed to vary the voltage amplitude. The shaft (8) is driven by an electromotor (7). The photo-sensor (5) registers a number of loading cycles. Adjusting the voltage supplied to the motor winding (7), a frequency

Table 2.1 Chemical composition of alloys (wt%)

	Si	Cu	Fe	Mn	Ti	Zr	Mg	Zn	Impurities
AK12	10–13	Up to 0.6	Up to 1.5	Up to 0.5	Up to 0.1	Up to 0.1	Up to 0.1	Up to 0.3	In total 2.7
	Fe	C	N	Si	O	H	Impurities	Ti	
VT1-0	0.25	Up to 0.07	Up to 0.04	Up to .01	Up to 0.2	Up to 0.010	Other 0.3	Balanced	

of motor rotation may be changed. In this study, researchers supplied a voltage of 90 V to the motor winding. A frequency of motor rotation was 15 Hz at this voltage.

2.3 Methods of Electron-Beam Processing

As outlined in the literature review, impulse electron-beam processing is currently one of the prospective methods to modify the subsurface in products with the aim to enhance their process characteristics [7, 13–20].

Compared to the widely applied laser technology, electron-beam processing makes it possible to control and vary an amount of supplied power, provides the target distribution of energy in the subsurface of a material being irradiated; additionally, it is distinguished by high efficiency. Institute of High Current Electronics, Siberian Branch, Russian Academy of Sciences (Tomsk, Russia) is a leader in the field of the development of impulse electron-beam devices with a plasma cathode.

A laboratory unit "SOLO" (Institute of High Current Electronics, Siberian Branch, Russian Academy of Sciences), the general view of which is shown in Fig. 2.3a, consists of an impulse electron source based on a grid-stabilized plasma cathode (Fig. 2.3b); a power supply unit of the electron source; a vacuum technological chamber with an inspection opening and a 2D moving table, where a plasma source of electrons and samples to be irradiated are placed; a control unit; a system for the checking of electron source and beam constants [14].

The above-mentioned unit has several serious advantages compared to early pulsed electron sources with a plasma cathode. These include a high-energy density in connection with a low accelerating voltage, a significant energy efficiency, a wide range of variable parameters, a good repeatability of impulses, a minimal preparation time, and a long service life (Table 2.3).

A beam of electrons is moved in an axial magnetic field provided that a pressure of argon in a vacuum chamber is in a range from 0.01 to 0.1 Pa. A space between magnet coils is 17 and 35 cm. Electrons emitted from a plasma edge and grid-stabilized enter a cumulative magnetic field, moving further in a magnetic field surrounding the coils. If the distance between coils is 17 cm, a generated electron beam will move toward a homogenous magnetic field, changing its diameter insignificantly (3–5%). Expanding the spacing between coils, an increase in the diameter of the electron

2.3 Methods of Electron-Beam Processing

Table 2.2 Chemical composition of alloy VT6 (wt%)

Fe	C	Si	V	N	Ti	Al	Zr	O	H	Impurities
Up to 0.6	Up to 0.1	Up to 0.1	3.5–5.3	Up to 0.05	86.45–90.9	5.3–6.8	Up to 0.3	Up to 0.2	Up to 0.015	Other 0.3

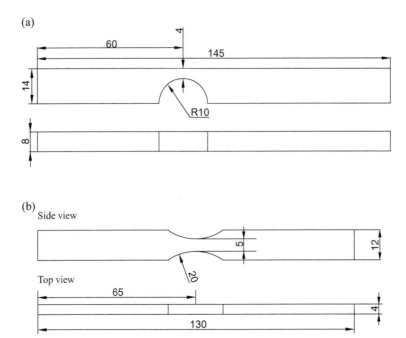

Fig. 2.1 A sample for fatigue tests: AK12 (**a**); VT1-0 (**b**) (unit: mm), reprinted from ref. [12] (Copyright 2018, with permission from Brazilian Metallurgical, Materials and Mining Association)

beam and its further refocusing are recorded. Therefore, the electron-beam diameter and energy density of a collector may be varied via selecting the correlation between a value of current in magnet coils and a distance to the collector (Fig. 2.4).

A narrow gap between magnet coils betters the distribution of current in the collector, making it more homogenous. A position of the collector in the diverging magnetic field (the space between magnetic coils set as 35 cm) increases a diameter of a beam reaching the collector.

Principal criteria accepted to characterize electron-beam processing of materials and determine temperature profiles of heated zones in surfaces, therefore, nature and kinetics of structure-phase transformations are suggested energy density in an electron beam as well as time and repetition rate of irradiation impulses.

2.3 Methods of Electron-Beam Processing

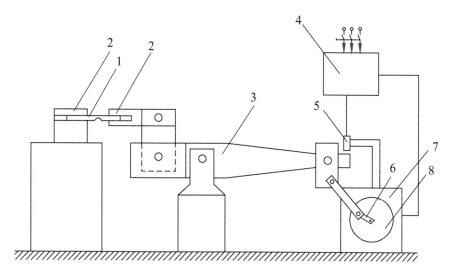

1 — sample; 2 — sample fixing device; 3 — lever mechanism; 4 — control unit: a frequency controller of a motor speed and a counter of cycles; 5 — photo-sensor; 6 — eccentric; 7 — motor; 8 — shaft

Fig. 2.2 Experimental unit for fatigue tests (1—sample; 2—sample fixing device; 3—lever mechanism; 4—control unit: a frequency controller of a motor speed and a counter of cycles; 5—photo-sensor; 6—eccentric; 7—motor; 8—shaft)

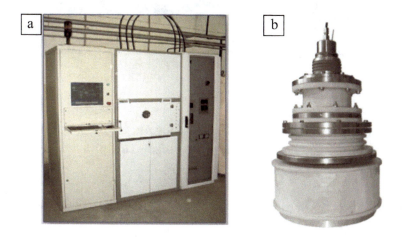

Fig. 2.3 A unit for the pulsed electron-beam surface modifying (**a**) and an electron source with a plasma emitter (**b**)

Table 2.3 Critical parameters of a vacuum impulse electron-beam unit "SOLO"

Parameter	Value
Overall dimensions/mm^3	1350 × 2150 × 2000
Overall dimensions of a vacuum chamber/mm^3	600 × 500 × 400
Current of a beam/A	20–250
Energy of electrons/keV	5–25
Energy density of electrons/(J/cm^2)	2–70
Impulse time/μs	50–200
Impulse repetition/Hz	0.3–20
Maximal consumed power/kW	2.5–10
Process pressure in a vacuum chamber/Pa	0.01–0.05
Beam diameter/mm	10–30
Zone of scanning/mm^2	200 × 200

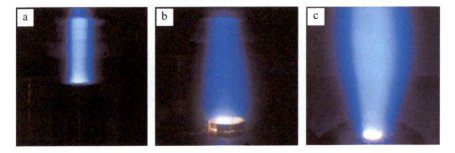

Fig. 2.4 An external view of a beam in the chamber. The collector is at a distance 10 (**a**), 14 (**b**), and 33 cm (**c**), respectively from a drift tube

2.4 Vacuum Impulse Electrical Explosion Apparatus EVU 60/10 for the Generation of Impulse Multiphase Plasma Jets

Electrical explosion alloying is a method to process a surface using concentrated energy flows. It is widely applied to modify the structure of various metals and alloys as well as to affect substantial and mechanical properties of a surface being irradiated. Face alloying involves the development of heterogeneous plasma flows from electrical explosion products; melting and saturation of the melt with alloying elements and further crystallization, which conduces to transformations of structure, physical and mechanical properties of a surface being treated.

2.4 Vacuum Impulse Electrical Explosion Apparatus EVU 60/10 …

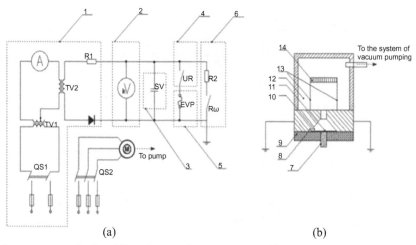

1 — power supply; 2 — kilo-voltmeter; 3 — energy capacitive storage; 4 — controlled dis-charger; 5 — impulse plasma accelerator; 6 — discharging circuit; 7 — inner cylindrical elec-trode; 8 — external ring electrode; 9 — insulator; 10 — discharge chamber; 11 — conductor; 12 — vacuum process chamber; 13 — sample clamps; 14 — a sample for spraying

Fig. 2.5 An electric circuit of a laboratory electrical explosion apparatus EVU 60/10 (**a**) and a diagram of a plasma accelerator (**b**) [22] (1—power supply; 2—kilo-voltmeter; 3—energy capacitive storage; 4—controlled discharger; 5—impulse plasma accelerator; 6—discharging circuit; 7—inner cylindrical electrode; 8—external ring electrode; 9—insulator; 10—discharge chamber; 11—conductor; 12—vacuum process chamber; 13—sample clamps; 14—a sample for spraying)

In some recent studies, researchers have formed heterogeneous plasma flows and investigated the effect of electron-beam processing on metals and alloys using an electrical alloying apparatus EVU 60/10 [21]. Figure 2.5 provides a functional electric circuit of the laboratory electrical explosion apparatus EVU 60/10. Its construction consists of three key components: a charger, a capacitive storage, and a plasma accelerator.

In a charger, there is an automatic current transformer, a step up transformer, and a rectifier. The apparatus runs in a semi-automatic mode. It is charged and discharged via pressing a correspondent digital key on a sensor display of a PLC-controller. The apparatus discharges automatically once a required amount of energy is accumulated in the capacitive storage. Overvoltage and malfunction protection of the unit allows discharging to a short-circuiting switch. The residual charge may be dumped by a beam for residual charge removal.

Behind the operation of a laboratory electrical explosion unit EVU 60/10 (Table 2.4), it is suggested to be the accumulating of energy by a battery of pulse capacitors and its following discharge over a conductor, which tends to fracture if a high density current flows through it.

The battery of the unit is mains-powdered over a time t_3 while an average consumed power is

Table 2.4 Principal characteristics of an electrical explosion apparatus EVU 60/10 M

Parameter	Value
Energy capacity/kJ	60
Discharge current frequency/kHz	10
Maximal charging voltage/kV	5
A setting increment of charging voltage/kV	0.1
Efficiency at a maximal charging voltage/(cycle/h)	10
Average consumed power at a charge less than/kW	0.55

$$P_3 = W/t_3$$

where $W = CU^2/2$—accumulated energy; C—total capacity of the battery; U—the value of charging voltage. The discharge to a conductor being exploded occurs at a time $t_p \ll t_3$ across an actuated arc gap; a switchboard of the charger is off. Here, a discharge power of the capacitive storage is generated in the plasma accelerator, being calculated as $P_н = \eta_p W/t_p$, where η_p is the efficiency of a discharge circuit. To ensure a multifold increase of a ratio $P_н/P_3$ impulse condensers are installed in the apparatus, which provide the supply of accumulated energy as short and strong impulses. The design of condensers allows their long-time operation in a mode similar to a short circuit. To reduce the inductance of a discharge circuit, they are connected in a battery by copper sheets. According to the unit specification, the maximal energy reserve of the capacitive storage is 60 kJ, the frequency of the discharge current is 10 kHz. The battery is charged at a voltage of 2–5 kV.

An arc gap consisting of two air-isolated flat electrodes has a low proper inductance and a long service life. The discharger is actuated once air is sucked out of the space between electrodes according to Paschen's law.

The impulse plasma accelerator is a coaxial end-type electrode system, which consists of inner and outer electrodes and an insulation spacer between them (Fig. 2.5b); a compression chamber with a directing nozzle; fastening clamps to fix a product being processed in front of an accelerator nozzle and place it in the process chamber. Low vacuum with residual pressure of 100 Pa is developed by a backing vacuum pump in the process chamber when charging the condenser battery.

Such parameters as impulse time, area of the surface being irradiated, and absorbed power density are taken into consideration when selecting a mode of electrical explosion alloying. The impulse time in the process of treatment is constant and set as 100 μs. An area of the surface to be hardened may be varied from 2–3 to 10–15 cm². A key parameter of the processing is the absorbed power density ranging from 1 to 8 GW/m². Diverse factors such as the charging voltage of a capacitive storage, diameters of an inner electrode in the plasma accelerator, and the nozzle of the process chamber as well as the distance from its edge to the surface being irradiated support this broad density range.

The products of conductor electrical explosion are revealed to be a multiphase system composed of a plasma component and condensed particles of various sizes. The components of a plasma jet build its front. Condensed particles have higher inertia and are concentrated in the rear of a jet. The quantity of plasma and condensed components of explosion products is not similar in the jet. The expanding of a jet brings about its further splitting into a fast plasma front and a relatively slow rear containing condensed particles. This is possible for smaller particles that have a higher speed. A dispersion degree of condensed particles is related to a value of charging voltage and a discharge time of the capacitive storage.

An average mass density of a jet approximates to 1 g/m^3 and contributes to a high degree of alloying in the process of treatment. A speed of a jet front is above 15 km/s. A speed of condensed particles in the rear of a jet amounts to several meters per second. As a result, a layer with high pressure and temperature is developed close to the surface being irradiated. The pressure varies from 5 to 15 MPa, and a temperature of plasma attains a level of 10^4 K. Selecting such parameters of a plasma jet, it is possible to melt a surface with the penetration of several tens of micrometers, support the degree of alloying up to a few tens of atomic percentage, and cool at a very high speed (10^5–10^7 K/s).

A coaxial end-type electrode system in the discharge chamber connected with a directing nozzle (Fig. 2.5b) raises a mass density of explosion products as well as a heat impact on the material till their values are sufficient to provide the alloying of a surface for a short impulse time (100 μs). Electrical and gas-dynamical forces make explosion products flow into the process chamber, where low vacuum is maintained. A product being processed is firmly fixed by a special clamp in front of the accelerator nozzle in the process chamber, where low vacuum with a residual pressure of 100 Pa is generated by a backing vacuum pump while the battery of condensers is charged.

In electrical explosion alloying, a protective layer composed of condensed particles of explosion products and weighted portions of powders is formed due to the effect of impulse multiphase plasma beams on the surface of metals and alloys. There are pores, craters, and micro-cracks detected in this coating, so a practical application of this method has specific limitations.

Principal physical parameters thought to be relevant for the alloying of a surface by concentrated energy flows and determining of a heat field in the material as well as an alloying degree are a specific power density q and an impact time τ. As soon as the value of q approximates to 10^3 W/cm^2, the temperature of the surface being irradiated equals to the temperature of melting, and provides favorable conditions for the alloying. Therefore, the impact time τ is typically short, being approximately 10^{-6}–10^{-3} s, i.e. the treatment has an impulse character. Electrical explosion alloying of metals and alloys is realized both by a plasma component of a jet, and by condensed particles. Here, the degree of alloying by a plasma component increases provided that q and p (pressure of a plasma jet on the surface) rise.

Once the charging voltage, heat, and pressure impacts on the surface scale up due to inhomogeneous pressure, the melt is forced out of the melting zone at a threshold value of voltage, i.e. its splashing is recorded. In fact, the surface ruptures in such treatment conditions. However, weighted portions of powder added to the jet eliminate a radial flow of the melt, making it possible to carry out the treatment without splashing in highly intensive modes. It may be connected with an increasing pressure of a jet on the edge of a nozzle; as a result, its gradient on a surface tends to drop. Another explanation is also suggested, i.e. if the melt of a base being treated fails to moisten powder particles, they can be concentrated on the surface, forming a layer, which hinders the entraining of the melt by a plasma flow.

2.5 Equipment for the Processing of Titanium Alloy Surface by Low-Energy High-Current Electron Beam

In the process of electrical explosion carbonization, numerous particles of carbon fibers penetrate the zone of alloying. When convective mixing of the melt is initiated by a plasma jet, they redistribute throughout the melt, reaching the boundaries of the molten zone. The treatment has an impulse character, so they are not capable to react completely with titanium and form titanium carbide. Therefore, it is possible to carry out an additional treatment of the alloying zone to dissolve particles of carbon fibers and raise a concentration of titanium carbide. A heat impact by low-energy high-current electron beams is recognized to be an efficient method applied for the modifying of structure and properties in metals and alloys. Continuous electron beams are used to provide the melting of a surface. A combination of electrical explosion alloying and electron-beam processing is efficient, since both methods demonstrate comparable diameters of treated zones (1–3 cm) and values of absorbed power, therefore, similar depths in the zone of melting.

The use of impulse-periodic electron beams allows the development of finely dispersive annealing structures; in addition, the repetition of impulses prolongs the time a surface is in a molten state.

In this study, the surface of VT6 alloy exposed to face alloying was processed using a laboratory unit "SOLO." The external view of this unit is presented in Fig. 2.3.

The above-mentioned unit "SOLO" consists of a control box, a power supply, a vacuum process chamber with a plasma source of electrons and a table to place a material for treatment on it. An essential characteristic of this unit is that an impulse electron beam is generated by an electron source with a plasma cathode. This feature made it possible for the first time to carry out an independent fine adjustment of principal parameters such as beam current, average energy of electrons, impulse time, and pulse repetition, varying them in a wide range and in different combinations. Parameters of an electron beam and a mode of processing are set up and controlled by computer.

Electron-beam processing is considered to be an advantageous technique for a number of reasons. They include:

1. An independent fine adjustment of main parameters in any combination.
2. Polishing and hardening of metals and various alloys without micro-cracks and micro-craters.
3. A 15-fold decrease in the roughness of a surface and the enhancement of surface hardness after treatment.

2.5 Equipment for the Processing of Titanium Alloy Surface ...

4. A possibility to process products with a complex physical configuration, e.g. dies, press dies.
5. It is an environmentally friendly process because of its implementation in vacuum.
6. No particular means of radiation protection are required.
7. Energy and resource efficiency compared with traditional methods of polishing, less cost-intensive polishing abrasive materials, high-speed polishing.

It is suggested that energy of electrons may be varied from 2 to 20 keV, a current force of a beam in a range from 20 to 200 A, a time of impulses may be set from 20 to 200 μs, their number from 2 to 200, and an impulse frequency from 0.3 to 20 Hz with the help of the unit [23, 24]. General sizes of the vacuum chamber are 600 mm × 500 mm × 400 mm, a zone of manipulator-assisted scanning is 200 mm × 200 mm. A beam of electrons is transported in a coaxial magnetic field when a pressure of a working gas (argon) in the vacuum chamber is set to be 0.01–0.05 Pa. If coils are located close to each other, the electrical current distributes better on the collector; whereas a position of the collector in the divergent magnetic field (a space between coils is widened up to 35 cm) allows expanding the diameter of a beam coming to the collector.

Such parameters as energy density of an electron beam, impact time, number and frequency of irradiating impulses are considered to be relevant for electron-beam processing of base materials, temperature profiles of heated zones in the surface, character, and kinetics of structure and phase transformations.

For the purpose of electrical explosion alloying, powders of boron, titanium diboride, boron carbide, and silicon carbide were used as a material to be exploded.

Processing conditions of alloy VT6. The carbonization of samples prepared from titanium alloy VT6 was conducted with the help of the explosion of carbon and graphite fibers (ТГН-2 М). Samples in a form of disks (5 mm thick and 20 mm in diameter) were processed by electrical explosion. Varying the charging voltage on the accelerator storage, a diameter of the nozzle channel, and a distance from its edge to a sample, a treatment mode, facilitating the process of liquid phase alloying, was selected. The values were set 2.2 kV, 20 and 20 mm, respectively. Here, a depth and a radius of the alloying zone were maximal; moreover, the melt was not forced out on the surface by a plasma jet. The data on absorbed power density on a jet axle, dynamical pressure in compression-stress state at the surface, impulse time and area of irradiated surface in samples of titanium alloy VT6 are given in Table 2.5.

Table 2.5 Parameters of electrical explosion alloying

Parameters of electrical explosion alloying	Value
Absorbed power density $q/(GW/m^2)$	5.5
Charging voltage/kV	2.2
Dynamical pressure p/MPa	8.5
Impulse impact time $\tau/\mu s$	100
Area of the surface S/cm^2	3

In the center, a penetration depth of alloying was 25 μm. In prior research [21, 25], such modes of electrical explosion alloying were classified as highly intensive. They produce a strong effect on treatment outcomes because of the pressure a jet produces on the surface. As a result, a radial flow of the melt from the center of a zone toward its edges, even splashing, is recorded. In addition, such treatment conditions provide maximal possible parameters such as a depth of the alloying zone, a degree of its saturation with alloying additives, and a level of properties to be formed. This is the reason that attention of researchers has been drawn to this method.

Further electron-beam processing of the surface was carried out using a laboratory unit "SOLO." When selecting modes of electron-beam processing, a great practical experience of the Department of Physics (Siberian Industrial University) in the field of a complex treatment combining electrical explosion alloying and electron-beam processing was taken into consideration. For instance, choosing a range of the surface energy density, it was important to melt as deep as possible a surface being irradiated without intense evaporation. The surface energy density depends on the energy of electrons forming a beam and the impulse time. The number of impulses is related to the period of treatment, in particular, to the time a surface layer is in a molten state.

The surface energy density, the time, and the number of impulses could be varied in a wide range; however, researchers selected a combination that was able to provide a maximal increase in the microhardness and depth of modified layers for AK12. They remained as follows: (i) surface energy density of an electron beam $E_S = 50$ J/cm^2, impulse time $\tau = 100$ μs, their quantity $N = 10$ impulses; (ii) $E_S = 60$ J/cm^2, $\tau = 100$ μs, $N = 10$ impulses. The energy of electrons was 18 keV, and the frequency of impulse repetition $f = 0.3$ Hz. For alloy VT1-0 the settings were: (i) surface energy density of an electron beam $E_S = 25$ J/cm^2, impulse time $\tau = 150$ μs, their quantity $N = 3$ impulses; (ii) $E_S = 30$ J/cm^2, $\tau = 150$ μs, $N = 3$ impulses. The energy of electrons was 16 keV, and the frequency of impulse repetition $f = 0.3$ Hz.

Irradiation was conducted in the environment of argon at a residual pressure of 0.02 Pa. Electron-beam processing was carried out in the inert (argon) environment in the process chamber at a pressure of 3.3×10^{-2} Pa. In these conditions an initially hardened surface was re-molten to a depth of 5–7 μm. To sum up, the processing of a surface in conditions described above resulted in heating and melting of the surface up to 10–20 μm and melt annealing at a speed of $\approx 10^6$ K/s.

2.6 Methods of Structural Studies

Such characteristics of the modified surface as structure, element and phase composition, defect substructure were researched using the methods of optical microscopy, scanning and transmission micro-diffraction electron microscopy, and X-ray analysis.

Metallographic studies and investigations into a chemical composition of alloys were carried out in a Resource Sharing Center "Materials Study" (Siberian State Industrial University). The preparation of probes for metallographic studies included

2.6 Methods of Structural Studies

the cutting of samples from a material processed by an electron beam using a diamond cutting-off wheel, grinding, polishing, and chemical etching. Before grinding, samples were fixed in holders and covered with an acrylic self-hardening material. Sandpaper with lowering abrasive-particle dispersion was used for polishing. The degree of surface polishing was checked using a microscope MBS (МБС)-9 at a magnification of 50–100×. This binocular microscope is distinguished with a good depth of focus and uses an inclined illumination, which creates a shadowing effect to detect details of the surface topography.

After being polished by sandpaper, samples were polished by fine grit diamond pastes. Thin sections were washed thoroughly before being examined by a microscope at a magnification up to 1000×. The polishing was completed as machining marks were removed and further treatment didn't increase the number and clearness of structural details on the surface. Samples of silumin were exposed to chemical etching in a water solution of a hydrofluoric acid (0.5 mL 40% HF + 100 mL H_2O) for 5 to 10 s [26].

For the purpose of metallographic investigations, researchers used an optical microscope Olympus GX71 equipped with a digital camera DP70 controlled by a software ImageScopeM. With the help of this instrument, it is possible to obtain an image of a small object at magnitudes up to 1000×. The optical microscope is used to determine a physical configuration, dimensions, structure, and other microstructure characteristics of metals and alloys [27, 28]. The chemical composition of silumin under consideration was analyzed by the methods of X-ray analysis using an X-ray fluorescent spectrometer Shimadzu XRF-1800 with a digital camera to control an area being researched.

To analyze the structure of the modified layer and the fatigue surface of silumin, a scanning electron microscope SEM-515 Philips was used; its principal parameters: accelerating voltage 0–3 kV, resolution 5 nm, 10–160,000× magnification. Recently, scanning electron microscopy (SEM) has found a broad application as a nondestructive test method of structure and element composition of the material surface.

SEM techniques employ a wide range of signals to generate while an electron beam reacts with the surface of a sample. As primary electrons penetrate into a sample, they dissipate and react with the object under study. An "exploring" electron beam contacts with an object being studied; as a result, scattered, secondary and Auger electrons are generated, in addition, it is reported on a current of absorbed electrons, X-ray and cathodoluminescent emission and passed electrons in case of a thin sample.

These methods are thought to be universal and provide a lot of valuable data; moreover, the equipment used for their realization is simple and easy to control. Unlike traditional light microscopy, the resolution and the depth of focus in SEM are high; the interpretation of images is easy due to their three-dimensional representation. If compared to scanning probe microscopy, in SEM it is possible to investigate larger zones of the surface, and to explore and analyze surface structures with rough topography (fractures, coarse coatings, etc.); it is possible to form images in a broad range of magnification, and to collect accurate data on the surface and subsurface layers [3].

On the images produced by a microscope, a scale appears automatically, making it easy to assess sizes of structural elements. Images were obtained both in backscattered and secondary electrons.

A region of signal application depends on the thickness of its generation layer in the material under study. The region of secondary electrons generation is up to 10 nm deep provided that the scattering zone is comparable with the diameter of an electron probe. Therefore, the resolution of an apparatus is maximal when imaging in secondary electrons (1–1.5 nm) and is limited principally by the diameter of an electron probe. A key parameter, determining the origination of secondary electrons, is an inclination angle of a surface spot under study (grazing angle of a beam of primary electrons). This effect is used to acquire data on the topography of a surface if specimens are explored in secondary electrons [29].

The region of scattered electrons generation is larger than that of secondary electrons generation. For instance, the depth varies from 0.1 to 1 µm, being related to a range of electrons in the material of a sample, i.e. it increases if the voltage accelerating primary electrons goes up, and the average atomic number of elements that make up the sample. Therefore, a section, producing a signal is far larger than that of a probe; as a result, the resolution of an apparatus is limited. The number of scattered electrons is dominated by the atomic number of an element and increases as the latter rises. These regularities are essential for the quality analysis of an element composition in the material surface. A spot on the material surface with a higher average atomic number reflects more electrons. Therefore, it is brighter than other zones of a sample on the monitor; as a consequence, we can estimate the number of phases in a material, analyze its microstructure without preliminary etching of a section. The contrast obtained is called a compositional contrast.

To sum up, a scanning electron microscope is developed to explore micro- and nano-structures of different surfaces. Using basic tools of scanning electron microscopes, it is possible to investigate the morphology of a surface, and measure sizes, orientation, and other parameters of micro- and nano-dimensional objects in a range from several centimeters to subnanometers and nanometers at a millionfold magnification as well as analyze local properties of their element composition (a maximal region of excitation 1 µm).

Characteristic X-ray emission of a crystal body has a discrete line spectrum and is thought to be a "passport" of an element. The energy of emission is proportional to the atomic number of a chemical element for X-ray in one region (Moseley's law). This fact allows carrying out both qualitative and quantitative research into the element composition of the material surface, i.e. X-ray microanalysis. To investigate the element composition of the material surface, X-ray microanalysis was conducted using a micro-analyzer EDAX ECON IV, which is a supplement to a scanning electron microscope Philips SEM 515.

2.6 Methods of Structural Studies

Besides electron-diffraction microscopy, X-ray phase analysis was used to research the phase composition of modified layers, i.e. qualitative and quantitative parameters of their phases, their concentration, dispersion, structure, and chemical composition (X-ray diffraction meter Shimadzu XRD 6000, imaging in copper filtered irradiation Cu-Kα1; monochromator CM-3121).

A scanning electron microscope is intended for the exploring of a real structure of thin sections in massive objects or powder, film, and other objects studied in Physics of Solids, Materials Science, Biology [14, 15, 30], e.g. metals and alloys, miscellaneous ceramics, carbon nano-materials (nano-tubes, fullerenes), nonorganic nano-powders and films, high-temperature superconductors, optoelectronic materials, etc.

Being an integrated component of up to date transmission electron microscopes, detectors of secondary and backscattered electrons make it possible to image a structure in secondary and backscattered electrons, using a compositional (according to the chemical element number) or topographical (based on the surface topography) contrasts.

The defect structure of fractured specimens was analyzed by TEM methods of thin foils (JEM-2100F, JEOL). A transmission electron microscope JEM-2100F has the following characteristics: a field-emission source (FEG), which produces a bright beam of electrons (hundred times brighter and more stable than an apparatus with a lanthanum hexaboride LaB6 cathode); point resolution—0.19 nm; line resolution—0.4 nm; a range of the accelerating voltage—80–200 kV; a magnification range from 50× to 1,500,000×; an electron-beam diameter in the TEM mode—2–5 nm.

Basic characteristics of an electron microscope JEM-2100F include high electro-optical parameters, diverse methods to obtain and display information, efficiency, and convenience of operation. A microscope is tooled with a bright LaB6 cathode, a digital scanning device, a mechanism to change the electron beam convergence angle to carry out investigations by a convergent beam, a goniometer with a piezo controller of the atomic-scale position of an object. A microscope construction is well-resistant to vibration. Images are displayed on a fluorescent monitor and on a monitor by a high resolution CCD camera with a big FOV.

Phase analysis of an object can be carried out by means of nano- and micro-beam diffraction. Objects may be imaged in an SEM mode, using a system of accumulating images and bright-field and dark-field detectors at an accelerating voltage 200 kV and high (atomic) resolution not less than 0.2 nm. A microscope is equipped with a system for element analysis conducted by the method of energy dispersive X-ray microscopy (INCAEnergyTEM 250 X-Max).

A TEM microscope JEM-2100F enables such types of material analysis:

- bright-field and dark-field transmission type imaging of thin objects;
- direct resolution imaging of a crystal lattice;
- scanning imaging;
- imaging in a convergent beam.

Using the methods of TEM, it is possible to determine the type and parameters of a crystal lattice in a matrix and secondary phases as well as the orientation correlation between a matrix and secondary phase particles, investigate the structure of grains and sub-grains, their crystallographic orientation and disorientation angles, and analyze the defect substructure of a material (find an occurrence level of defects in the crystal structure, explore the density and distribution of dislocations, a type of a developing substructure, research inner stress fields, etc.). Such a set of measurable parameters allows collecting reliable data on hardening and fracturing mechanisms, finding out the importance of alternating phase composition and defect sub-structure for these processes.

Despite substantial advantages of TEM, this method of research still suffers from certain limitations. The latter include a necessity to produce enough vacuum to obtain a reliably high resolution, a lacking possibility to analyze big samples, the development of an atomic resolution in critical conditions for the material, i.e. the energy of an exploring electrons beam ranging up to 300 kV. Since a principle domain of TEM methods is the analysis of thin (a thickness below 200–300 nm) materials, sample preparation for this research is to be carried out accurately. Objects under study are divided into two types, assisted and self-sustaining samples. The first type of samples (mostly powders) are placed on membranes (thin foils), which are fastened to special grids. The grids with various openings are made of copper and rarely of nickel, molybdenum, tantalum, or gold.

Foils to hold the powder are made of graphite (vacuum spraying of carbon rods), and special polymers. Self-sustaining samples are produced in several phases. At the beginning, plates of certain physical configuration and thickness (100–200 μm) are cut of a massive sample. Then they are grinded and polished mechanically, improving defects caused by cutting. Several procedures to thin a cut plate (a required thickness is 100–200 nm) are available. For instance, electrolytic etching is used to produce foils of conductors (metals and alloys). This method is quite fast—from some minutes to one hour—and there is no risk of mechanical damages. However, chemical characteristics of the surface in the object under consideration can be changed. In addition, it might be harmful for human health. Foils of ceramic materials are made by ion etching, i.e. the thinning of a sample is possible due to the attacking of its surface by ions and neutral atoms. The conditions of thinning depend on the accelerating voltage, the type and current of ions, geometry (a grazing angle of electrons onto a sample), the temperature of a sample (the cooling in liquid nitrogen), etc. Ion etching is considered to be a universal thinning method, although its application is limited. Critical drawbacks restricting its usage are as follows. Ion etching may cause radiation-related damages of a material, the precipitation of a material being sprayed is possible, i.e. a transfer of a material from a foil section onto another one, and in addition, grooves may be formed along the path of an ion beam (a sample is rotated during spraying to minimize this defect).

Electrolytic etching of samples to prepare foils for the further TEM-based exploration of silumin structure and phase state failed for coarse inclusions of silicon fell out of a thinned plate, therefore, the samples were made by ion etching of plates ($h \approx 100$ μm) cut of a sample using the electro spark method (Fig. 2.6). The mode of cutting was selected carefully to avoid unnecessary deformation, so it has no effect on the structure of a sample. Foils for the analysis were cut in certain locations. This

2.6 Methods of Structural Studies

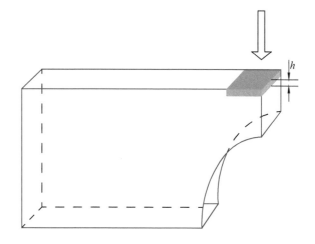

Fig. 2.6 A preparation diagram of an AK12 sample fractured when fatigue loading to prepare foils for electron-microscopic research, a curly arrow indicates a face of the sample

is a maximally stressed zone (fracture surface) in the center of samples. Thin zones (areas suitable for electron-microscopic analysis) were at a minimal distance from the fracture surface of a sample.

The thinning of cut plates was conducted via ion thinning (Ion Slicer EM-09100IS). A distinctive feature of this device is that no preparation of a disk thinned in the center is required. Samples for Ion Slicer are prepared gradually: a parallelepiped 2.8 mm × 0.5 mm × 0.1 mm is produced, and its broad face is covered by a special protective material and thinned by an argon ion beam. The energy of a beam is set below 8 kV, and the inclination angle is variable in a range from 0° to 6° with reference to the bigger surface of a sample. This fact allows minimizing radiation damages, therefore, preserving the initial structure and phase composition of a sample, which are analyzed further by SEM.

Bright-field images of a thin structure were used to classify morphological characteristics of the structure, and determine sizes, volume ratio, and localization spots of phases. To identify new phases, researchers used a method specified below. Once bright-field images of the material structure had been obtained, particles of the second phase were explored. For this purpose, two or three dark-field images were produced in various reflexes of these particles, marking reflexes of crystal lattice of these particles. Having shot a micro-electron-diffraction pattern of a foil zone to be analyzed, researchers measured radius vectors of reflexes in a particle being studied as well as angles between them. Further, a reciprocal lattice was developed with reflexes of a particle being analyzed. The phase identification involved then a search for such a reciprocal lattice, the section of which could be presented by this spot electron-diffraction pattern. Researchers explored reciprocal lattices of the most probable phases for this type of electron-beam processing. Additionally, available phase diagrams of balanced systems were used.

One of the most precise and quick-response methods, i.e. the measurement of microhardness was employed to analyze physical and mechanical properties of the surfaces. The difference in microhardness before and after processing might indicate the hardening of modified surface layers in metals and alloys. For the purpose of this research, microhardness was measured via Vickers indenter of a unit HVS-1000A with digital display—readings shown in hardness values HV. A load of an indenter was set 0.098 N. Values of microhardness were calculated according to measured diagonals of several (mostly 3–5) samples. The averaging was carried out using measurement data of several (3–5) samples. The achieved accuracy ranged from 7 to 10%. The behavior of hardness in the modified layer was analyzed via the assessing of microhardness in the irradiated surface and developing microhardness vs. depth profiles.

Moreover, strength properties of the material were tested according to nano-hardness (NANO Hardness Tester NHT-S-AX-000X), with a load of the indenter varying from 5 to 300 mN (nano-indention).

The term "nano-indention" refers to a group of methods using a local precise impact on the material with the simultaneous registration of deformation responses with nano-meter resolution. The matter of these methods is a programmed application of small and ultra-small forces to an indenter and simultaneous recording of a ratio resistance force P vs. displacement (depth of indenting h). Nano-indenting and nano-hardness are considered for a situation when a response to a local load is caused by a small plastic deformation when contrasted with high elastic deformation ($\approx 75\%$). In micro-indenting, plastic deformation develops in crystal materials, realizing due to originating and displacing dislocations. It requires a force varying from tens to thousands of mN, that complies with an indenter penetration depth ≈ 0.1–10 μm [31].

Tribological properties (wear resistance and friction coefficient) were studied in the geometry disk-rod using a high-temperature tribometer (CSEM, Switzerland) at room temperature and humidity. As a counter-body a hard alloy BK-8 ball with a diameter of 3 mm was used, and the track diameter was set to be 6 mm, the speed of rotation 2.5 cm/s, load 3 N, the distance to stop 38.6 m, the number of turns 3000. A wear resistance criterion appears to be a specific volume of material wear track, which was measured by a laser optical profile measuring device MicroMeasure 3D Station (Stil, France) and calculated according to the formula:

$$V = \frac{2 \cdot \pi \cdot R \cdot A}{F \cdot L}$$

where R—the radius of track; A—the cross-section area of the wear groove; F—the applied force; L—the distance of a ball.

A quantitative analysis of the steel structure was conducted using stereology methods and quantitative electron microscopy; a phase analysis of steel was carried out by indenting of micro-electron-diffraction patterns applying a dark-field method.

2.7 Methods of Quantity-Related Proceeding of Research Data

Estimation of mean grain dimensions. Mean grain dimensions were estimated by the random linear intercept method applied to micro-sections. Grain boundaries were etched in an electrolyte selected individually for each material. The mean size of grains (D) in the material volume was estimated in view of mean grain sizes measured in micro-sections:

$$\bar{D} = 0.5\pi \left(\bar{d}^{-1}\right)^{-1} \quad (2.1)$$

where $\bar{d}$—the mean grain size determined according to a micro-section:

$$\bar{d}^{-1} = N^{-1} \sum_{i=1}^{N} d_i^{-1} \quad (2.2)$$

where N—the number of measurements; d_i—the actual grain size on a micro-section. The mean root square deviation (σ_D) was calculated according to the formula:

$$\sigma_D = \sqrt{4/\pi \left(\bar{d} \cdot \bar{D}\right) - \left(\bar{D}\right)^2} \quad (2.3)$$

Estimation of a dislocation sub-structure volume percent (P_V). Koneva N. A. et al. have applied this method to explore dislocation sub-structures developing in one-phase alloys under deformation. Since dimensions of structural elements in the forming dislocation sub-structures are bigger or similar to a foil size, their imprints in foil might be considered as random cross-sections in a metallographic sample. Therefore, the volume percent was determined in view of random cross-sections, i.e. the percent of the foil area Ps occupied by a certain type of the dislocation sub-structure was estimated, in other words, researchers used a planimetric method. Within this approach, image areas of each dislocation sub-structure on the surface to be analyzed were calculated. Then obtained data were summed up. The sum was divided by the area of the surface spot being analyzed.

In an isotropic structure, P_V might be determined on one representative cross-section of a crystal. In a heterogeneous structure, a representative selection is to be carried out on some differently oriented cross-sections.

Estimation of a scalar density of dislocations. The scalar density of dislocations was estimated by the secant method adjusted by the invisibility of dislocations. A rectangular grid was used as a test line. The scalar density of dislocations on micro-electron-diffraction patterns taken in electron-microscopic studies is calculated according to the formula:

$$\rho = \frac{M}{t}\left(\frac{n_1}{l_1} + \frac{n_2}{l_2}\right) \quad (2.4)$$

where M—the magnification of a micro-diffraction pattern; n_1 and n_2—the crossing number of dislocations and horizontal l_1 and vertical l_2 lines, respectively (l_1 and l_2—a sum length of horizontal and vertical lines).

The scalar density of dislocations was estimated for each particular type of a dislocation substructure. A mean value of the scalar density was calculated using the volume percent of all dislocation substructures according to the formula:

$$\rho = \sum_{i=1}^{Z} P_{V_i} \rho_i \qquad (2.5)$$

where ρ_i—the scalar density of dislocations in a particular dislocation sub-structure; P_{V_i}—the volume percent of the material occupied by this dislocation sub-structure; Z—the number of dislocation sub-structures.

References

1. Cui, W.: A state-of-the-art review on fatigue life prediction methods for metal structures. J. Mar. Sci. Technol. **7**, 43–56 (2002)
2. Stephens, R.I., Fatemi, A., Stephens, R.R., Fuchs, H.O.: Metal Fatigue in Engineering (2000)
3. Gromov, V.E., Ivanov, Y.F., Vorobiev, S.V., Konovalov, S.V.: Fatigue of steels modified by high intensity electron beams. In Fatigue of Steels Modified by High Intensity Electron Beams, pp. 1–308. Cambridge International Science Publishing, London (2015)
4. Gromov, V.E., et al.: Increase in fatigue life of steels by electron-beam processing. J. Surf. Investig. X-ray, Synchrotron Neutron Tech. **10**, 83–87 (2016)
5. Sizov, V.V., Gromov, V.E., Ivanov, Y.F., Vorob'ev, S.V., Konovalov, S.V.: Fatigue failure of stainless steel after electron-beam treatment. Steel Transl. **42**, 486–488 (2012)
6. Vorob'ev, S.V., Gromov, V.E., Ivanov, Y.F., Sizov, V.V., Sofroshenkov, A.F.: Nanocrystalline structure and fatigue life of stainless steel. Steel Transl. **42**, 316–318 (2012)
7. Gromov, V.E., Gorbunov, S.V., Ivanov, Y.F., Vorobiev, S.V., Konovalov, S.V.: Formation of surface gradient structural-phase states under electron-beam treatment of stainless steel. J. Surf. Investig. X-ray, Synchrotron Neutron Tech. **5**, 974–978 (2011)
8. Ivanov, Y.F., Gromov, V.E., Gorbunov, S.V., Vorob'ev, S.V., Konovalov, S.V.: Gradient structural phase states formed in steel 08Kh18N10T in the course of high-cycle fatigue to failure. Phys. Met. Metallogr. **112**, 81–89 (2011)
9. Grishunin, V.A., Gromov, V.E., Ivanov, Y.F., Volkov, K.V., Konovalov, S.V.: Evolution of the phase composition and defect substructure in the surface layer of rail steel under fatigue. Steel Transl. **43**, 724–727 (2013)
10. Gromova, A.V., et al.: Ways of the dislocation substructure evolution in austenite steel under low and multicycle fatigue. In Procedia Engineering. (2010). https://doi.org/10.1016/j.proeng.2010.03.009
11. Gromov, V.E. et al.: Steel fatigue life extension by pulsed electron beam irradiation. J. Surf. Investig. X-ray, Synchrotron Neutron Tech. **9**, 599–603 (2015)
12. Konovalov, S., Komissarova, I., Ivanov, Y., Gromov, V., Kosinov, D.: Structural and phase changes under electropulse treatment of fatigue-loaded titanium alloy VT1-0. J. Mater. Res. Technol. **8**, 1300–1307 (2019)
13. Engelko, V., Yatsenko, B., Mueller, G., Bluhm, H.: Pulsed electron beam facility (GESA) for surface treatment of materials. Vacuum **62**, 211–216 (2001)

14. Ozur, G.E., Proskurovsky, D.I., Rotshtein, V.P., Markov, A.B.: Production and application of low-energy, high-current electron beams. Laser Part. Beams **21**, 157–174 (2003)
15. Rotshtein, V.: Microstructure of the near-surface layers of austenitic stainless steels irradiated with a low-energy, high-current electron beam. Surf. Coatings Technol. **180–181**, 382–386 (2004)
16. Gromov, V.E., Ivanov, Y.F., Sizov, V.V., Vorob'ev, S.V., Konovalov, S.V.: Increase in the fatigue durability of stainless steel by electron-beam surface treatment. J. Surf. Investig. X-ray, Synchrotron Neutron Tech. **7**, 94–98 (2013)
17. Ivanov, Y.F., et al.: Multicyclic fatigue of stainless steel treated by a high-intensity electron beam: surface layer structure. Russ. Phys. J. **54**, 575–583 (2011)
18. Gromov, V.E., Konovalov, S.V., Aksenova, K.V., Kobzareva, T.Y.: The Evolution of the Structure and Properties of Light Alloys at the Energy Influences. Publishing house of the Siberian Branch of the Russian Academy of Sciences, Novosibirsk (2016)
19. Ivanov, Y.F., et al.: Pulsed electron-beam treatment of WC–TiC–Co hard-alloy cutting tools: wear resistance and microstructural evolution. Surf. Coatings Technol. **125**, 251–256 (2000)
20. Proskurovsky, D.I., Rotshtein, V.P., Ozur, G.E., Ivanov, Y.F., Markov, A.B.: Physical foundations for surface treatment of materials with low energy, high current electron beams. Surf. Coatings Technol. **125**, 49–56 (2000)
21. Romanov, D.A., Budovskikh, E.A., Zhmakin, Y.D., Gromov, V.E.: Surface modification by the EVU 60/10 electroexplosive system. Steel Transl. **41**, 464–468 (2011)
22. Ivanov, Y. F. et al. On the fatigue strength of grade 20Cr13 hardened steel modified by an electron beam. J. Surf. Investig. X-ray, Synchrotron Neutron Tech. **7**, 90–93 (2013)
23. Chudina, O.V., Petrova, L.G., Borovskaya, T.M.: Mechanisms of hardening of iron by laser alloying and nitriding. Met. Sci. Heat Treat. **44**, 154–159 (2002)
24. Chudina, O.V., Petrova, L.G., Borovskaya, T.M.: The hardening mechanisms of iron upon laser alloying and nitriding. Metalloved. i Termicheskaya Obrab. Met. 22–23 (2002)
25. Tsvirkun, O.A., Bagautdinov, A.Y., Ivanov, Y.F., Budovskikh, E.A., Gromov, V.E.: Phase composition and defect substructure of nickel alloyed with boron and copper by electric explosion of conductors. Russ. Phys. J. **50**, 199–203 (2007)
26. Yang, S., et al.: Surface microstructures and high-temperature high-pressure corrosion behavior of N18 zirconium alloy induced by high current pulsed electron beam irradiation. Appl. Surf. Sci. **484**, 453–460 (2019)
27. Murphy, D.B., Davidson, M.W.: Light Microscopy. Fundamentals of Light Microscopy and Electronic Imaging, Vol. 689. Humana Press, Clifton (2011)
28. Murphy, D.B., Davidson, M.W.: Fundamentals of Light Microscopy. in Fundamentals of Light Microscopy and Electronic Imaging, pp. 1–19. Wiley, New York (2012). https://doi.org/10.1002/9781118382905.ch1
29. Paredes, A.M.: Microscopy | Scanning electron microscopy. In Encyclopedia of Food Microbiology, pp. 693–701. Elsevier, Amsterdam (2014). https://doi.org/10.1016/b978-0-12-384730-0.00215-9
30. Konovalov, S., Ivanov, Y., Gromov, V., Panchenko, I.: Fatigue-induced evolution of AISI 310S steel microstructure after electron beam treatment. Materials **13**(20), 1–13 (2020)
31. Golovin, Y.I.: Nanoindentation and mechanical properties of solids in submicrovolumes, thin near-surface layers, and films: a review. Phys. Solid State **50**, 2205–2236 (2008)

Chapter 3
Structure and Properties of As-Cast Silumin and Processed by Intense Pulsed Electron Beam

As mentioned in Chapter 2, commercial as-cast silicon-aluminum alloy (silumin) AK12 is used for the purpose of research. Its element composition according to GOST 1583—93 (Russian State Standard) is provided in Table 2.1.

Silumin compounds containing iron, copper, manganese, and other alloying elements, are thought to be prospective up to date light-weight materials. They are distinguished by sufficient wear resistance, castability, and thermal stability. However, these alloys suffer from certain limitations, e.g. poor pressure shaping because of a tendency to cracking. Alloying elements (iron, manganese, copper, etc.) enhance strength, on one hand, but worsen resistance to cracking of silumin, on the other hand, forming intermetallic compounds with lamellar morphology.

Resource properties of the material may be improved significantly if the results-oriented development of additional structure-phase state levels is realized in sub-micro and nano-scale ranges. An efficient method of such modification is the processing of a material surface by an intense sub-millisecond pulsed electron beam, which changes the structure of a surface layer (up to tens of micrometers); as a consequence of this process, the surface is in a multimodal structure-phase state, whereas the structure-phase state of the main alloy volume is practically stable.

This chapter summarizes data obtained when certifying the structure of as-cast silumin AK12 and modified by an intense pulsed electron beam. To research the structure and phase composition, methods of optical (etched micro-sections) microscopy, SEM and TEM were used. For the purpose of element study, X-ray microanalysis was carried out. The phase composition was assessed using the method of X-ray phase analysis.

3.1 Structure-Phase Study of As-Cast Silumin

As outlined above, properties of silumin depend basically on dimensions, morphology, and a positional relationship of the second phase inclusions (mainly

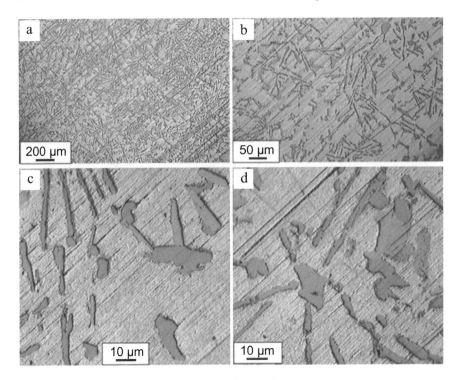

Fig. 3.1 Structure of an etched as-cast silumin micro-section

intermetallic compounds). This section aims to analyze the morphology, element, and phase composition of inclusions in as-cast silumin AK12.

Figure 3.1 presents a structure image of an etched as-cast silumin AK12 microsection. As seen, silumin under consideration is a multiphase material. The most considerable part of the second phase inclusions is lamellar. Besides lamellae, there are polygon and irregular inclusions.

Metallographic research conducted by the methods of selective etching has indicated several types of inclusions in silumin.

1. Lamellar light-grey inclusions—phase β (Al_5SiFe);
2. Brown inclusions in a form of regular polygons—phase α ($Al_{15}(FeMn)_3Si_2$); particles looking like Chinese hieroglyphs if a concentration of iron is low;
3. Oval grey inclusions—particles of silicon.

Figure 3.2 provides an image of as-cast silumin AK12 structure shot by the methods of SEM in backscattered electrodes. Clearly, image contrasts of second phase inclusions differ significantly in intensity. The most substantial share of inclusions appears to have moderate intensity, which is nearly similar to the image of a matrix. Inclusions formed by atoms with a higher atomic mass are suggested to reflect an upcoming flow of electrons more intensively; therefore, their contrast is lighter in images of the structure shot in backscattered electrons.

3.1 Structure-Phase Study of As-Cast Silumin

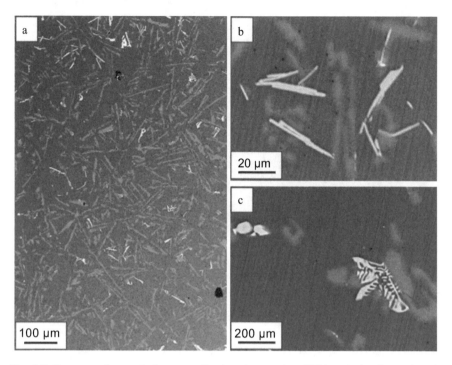

Fig. 3.2 Structure of an etched as-cast silumin micro-section. SEM analysis of a surface in backscattered electrons

This fact allows concluding that the element composition of inclusions found in the alloy varies largely. The most inclusions detected in silumin are lamellar, saturated with silicon atoms having intensity similar to the matrix image (Fig. 3.2a). Needle-shaped inclusions (Fig. 3.2b) and inclusions looking like Chinese hieroglyphs (Fig. 3.2c) are saturated with elements, the atomic mass of which exceeds that of aluminum.

A quantitative investigation into the element composition of as-cast silumin AK12 was conducted using the method of X-ray microanalysis. An averaged element composition of a local material zone was determined via assessing the surface of a micro-section. With this purpose, a polished surface of zones (1450 μm × 1050 μm^2) was examined with the help of X-ray microanalysis; the thickness of the layer under study was ≈ 5 μm (Fig. 3.3). The data presented in Table 3.1 indicate quite expectably that aluminum and silicon remain core elements in the zone being analyzed. The concentration of silicon on average (according to the data of 6 microzones) amounts to 21.5 wt% (Table 3.1). Since no lines of elements heavier than aluminum and silicon were found on the energy spectra, their volume percentage was assumed to be low (1–2 wt%). To clarify this question, X-ray microanalysis of several inclusions was carried out (Fig. 3.4).

Three types of particles with different image intensity in backscattered electrons and morphology (see Fig. 3.3) were selected for the element analysis. The data given

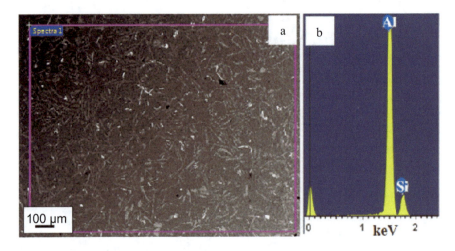

Fig. 3.3 Image of aluminum–silicon alloy surface shot in X-ray microanalysis (**a**) and energy spectrum obtained on the zone (**a**) of the alloy surface (**b**)

Table 3.1 X-ray microanalysis data of as-cast silumin AK12 (unit = wt%)

Element	Zone						Mean
	1	2	3	4	5	6	
Al, K	78	77	79	78	80	79	78.5
Si, K	22	23	21	22	20	21	21.5

Fig. 3.4 Image of aluminum–silicon alloy surface shot in X-ray microanalysis; zones of element analysis indicated

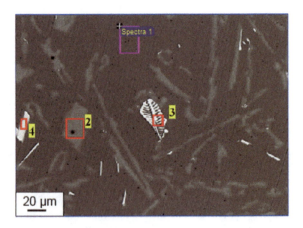

in Table 3.2 suggest that besides aluminum and silicon, iron and manganese are found in the alloy under consideration; furthermore, their distribution pattern differs significantly throughout the material. The matrix (Fig. 3.4, Spectrum 1) is composed of aluminum atoms (within minimum detectable limits). Particles with a reasonably

3.1 Structure-Phase Study of As-Cast Silumin

Table 3.2 X-ray microanalysis data of the aluminum–silicon alloy surface given in Fig. 3.4 (unit = wt%)

Element	Zone			
	1	2	3	4
Al	99	2.0	57	54
Si	1.0	98	8	15
Mn	0.0	0.0	1	1
Fe	0.0	0.0	34	30

low contrast level seen in the images structure obtained in backscattered electrons are composed of silicon atoms (Fig. 3.4, Spectrum 2). Particles with a brighter phase contrast contain aluminum, silicon, iron, and manganese atoms (Fig. 3.4, Spectrums 3 and 4).

The data obtained demonstrate that the concentration of iron and silicon atoms is important for the morphology of particles. For instance, a ratio of concentrations in particles looking like skeletons (Fig. 3.4, Spectrum 3) is Al/Fe/Si = 7/4/1; that in particles with lamellar morphology (Fig. 3.4, Spectrum 4) is Al/Fe/Si = 3/2/1. According to the data presented in studies [1, 2], it is assumed that particles with lamellar morphology (Fig. 3.4, Spectrum 4) are of β-phase (Al$_5$(Fe,Mn)Si); skeleton-shaped particles (Fig. 3.4, Spectrum 3) represent α-phase (Al$_8$(Fe,Mn)$_2$Si).

The phase composition of as-cast silumin AK12 was investigated in X-ray structure analysis. A zone of an X-ray pattern obtained on the material under study is presented in Fig. 3.5. Table 3.3 shows the analysis findings of a micro-electron-diffraction pattern.

The data presented in Table 3.3 suggest the sufficient compatibility of X-ray microspectroscopy (SEM) and X-ray structure analysis results. These include similar ratios of aluminum and silicon phases, lacking lines of intermetallic phases in X-ray patterns; as mentioned above, these facts indicate a low volume percentage of these inclusions. Moreover, parameters of aluminum and silicon crystal lattices in the alloy under consideration are similar to those of pure elements, confirming the full unmixing of these elements in the course of the alloy crystallization.

However, the data of element (Table 3.1) and phase (Table 3.3) compositions obtained when analyzing a certain sample differ from the concentration of silicon in industrial silumin AK12. As specified in GOST 1583—93, there is to be 10–13 wt% silicon in silumin AK12. To answer this question, supplementary research into the chemical composition of silumin was carried out employing the methods of X-ray spectroscopy. The outcomes are provided in Table 3.4.

The data in Table 3.4 demonstrate that X-ray spectroscopy allows investigating bigger volumes of material in comparison with X-ray microanalysis and has revealed that the concentration of silicon in silumin under consideration is similar to the alloy grade. According to the data presented in Tables 3.1, 3.3, and 3.4 and for the purpose of further discussion, the alloy used in this research is suggested to be silumin AK12, a distinctive feature of which is significant inhomogeneity of silicon pattern in the ingot.

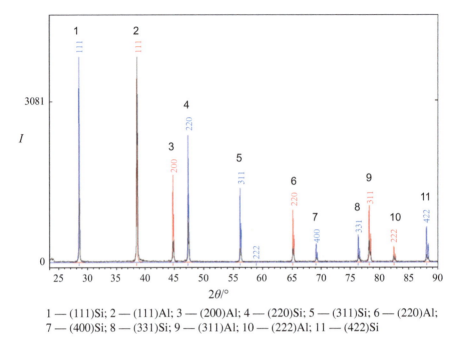

1 — (111)Si; 2 — (111)Al; 3 — (200)Al; 4 — (220)Si; 5 — (311)Si; 6 — (220)Al; 7 — (400)Si; 8 — (331)Si; 9 — (311)Al; 10 — (222)Al; 11 — (422)Si

Fig. 3.5 Fragment of an Al–Si alloy X-ray pattern. Numbers indicate diffraction maximums

Table 3.3 The X-ray structure analysis data of as-cast silumin

Phase	Concentration/wt%	Lattice type	Lattice parameter/nm		Atom radius/nm
			a_0	a	
Al	76.5	Fm3m	0.40500	0.40514	0.143
Si	23.5	Fm3ms	0.54307	0.54344	0.132

Note a_0—reference value; a—value in the alloy

Table 3.4 X-ray spectroscopy results of as-cast silumin AK12 (mass percentage) (unit = wt%)

Si	Fe	Cu	Mn	Zn	Mg	Ni	Ti	Cr	Al
8.99	0.38	0	0.022	0	0.058	0.004	0.008	0.002	Balanced

3.2 Structure and Phase Composition of Silumin Irradiated by an Intense Pulsed Electron Beam

As stated above, the processing of a material surface by an intense sub-millisecond electron beam represents an efficient method to modify the surface of a material. This section aims to analyze principles dominating the modification of silumin AK12 structure and phase composition by an intense pulsed electron beam.

3.2 Structure and Phase Composition of Silumin Irradiated ... 81

The prior research has pointed out that the irradiation of aluminum alloy (12.49% Si, 2.36% Mg, 0.6% Cu, 0.35% Ni, 0.3% Fe, at.%) surface by an intense electron beam (experimental unit "SOLO," Institute of High Current Electronics, Siberian Branch, Russian Academy of Sciences) results in significant transformations of the material structure, i.e. a two-phase alloy composed of high-speed crystallization aluminum cells (sizes range within 1 μm) is formed in a surface layer (up to 50 μm). Cells are surrounded by thin (up to 100 nm) silicon layers. Mechanical tests of modified samples have demonstrated that the hardness of the surface layer ($\approx$ 50 μm) is $\approx$ 1.5 fold higher than that of the main material volume.

The surface of as-cast silumin AK12 was processed by an intense pulsed electron beam. Samples prepared for fatigue tests were irradiated by operating an experimental unit "SOLO" [3] in the conditions: energy of electrons 18 keV; impulse repetition 0.3 s; pulse duration of electron beams 50 μs and 150 μs; energy density of an electron beam 10, 15, 20, 25 J/cm^2; number of impulses 1, 3, 5. The surface modified by an electron beam was analyzed by the methods of SEM and optical microscopy.

It was mentioned above that there are a lot of lamellar silicon crystallites in as-cast silumin AK12 (Fig. 3.6). Lamellae are located chaotically or they decorate alloy grain boundaries. The lamellae of silumin under study in a micro-section plane vary in a range from one to tens of micrometers. Undoubtedly, a material with such an extensive number of brittle inclusions with diverse forms and sizes has low resistance to cracking.

Optical microscopy of the irradiated surface has revealed the following outcomes. The morphology of silicon wasn't changed (Fig. 3.7) when the silumin surface was irradiated by an electron beam (10 J/cm^2, 50 μs, 5 impulses). Similar to the initial state, silicon is in a form of elongated lamella. Representing a characteristic of the structure, the fragmentation of silicon lamellae might be a result of elastic stresses developing in the material due to high cooling speeds of the impulse treatment (Fig. 3.7c, d).

The lamellae are divided into zones both lengthwise and transversally. Longitudinal sizes vary within 5 μm, and the range of transverse sizes is 2 μm. Therefore,

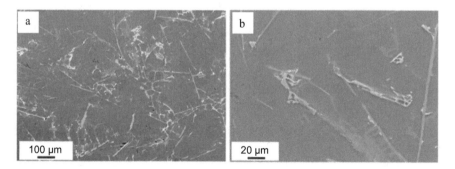

Fig. 3.6 Structure of the etched as-cast silumin micro-section

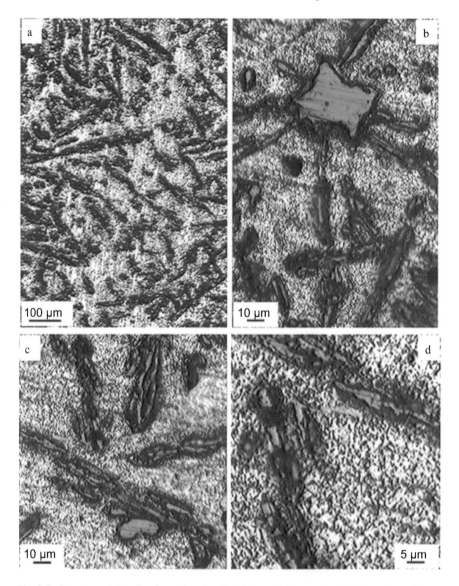

Fig. 3.7 Structure of the silumin surface irradiated by an intense pulsed electron beam (energy density of a beam 10 J/cm^2, 50 μs, 5 impulses)

a combination of short lamellae packages replaces silicon lamellae in the material when silumin is irradiated by an electron beam (10 J/cm^2, 50 μs, 5 impulses).

It is reported on the globularization of silicon inclusions (Fig. 3.8) when the surface of silumin is irradiated by an electron beam (15 J/cm^2, 150 μs, 3 impulses, i.e. at higher densities of energy). Besides fragmented silicon lamellae, similar to lamellae found in a sample irradiated by an electron beam (10 J/cm^2, 50 μs, 5 impulses

3.2 Structure and Phase Composition of Silumin Irradiated …

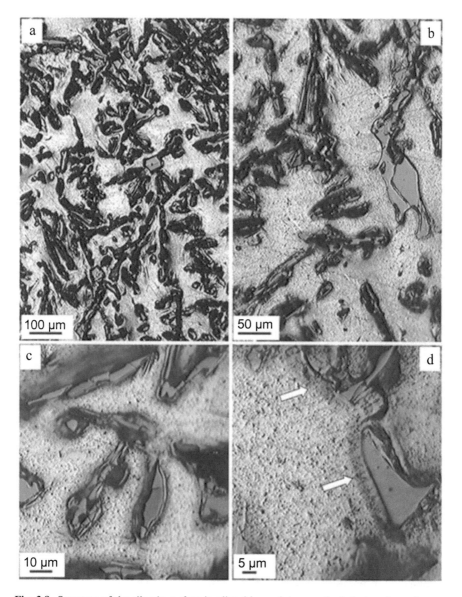

Fig. 3.8 Structure of the silumin surface irradiated by an intense pulsed electron beam (energy density 15 J/cm^2, 150 μs, 3 impulses). Arrows (d) indicate layers forming along the silicon-aluminum boundary

[Fig. 3.7]), there are globular silicon inclusions in the surface of irradiated samples; longitudinal sizes vary from 10 to 20 μm, transverse ones are in a range from 3 to 8 μm (Fig. 3.8a, b).

The second distinctive feature of the silumin surface revealed in the irradiation by an electron beam (15 J/cm^2, 150 μs, 3 impulses) represents the development of elongated layers with a thickness up to 5 μm (Fig. 3.8d, layers indicated by arrows) along the lamella-matrix boundary. Apparently, this fact confirms the dissolution of silicon inclusions alongside the mechanical fracture in the surface layer in the conditions of irradiation above.

When the surface of silumin is irradiated by an electron beam (20 J/cm^2, 150 μs, 5 impulses), a structure differing enormously from the initial one develops in the surface. As seen in Figs. 3.9 and 3.10, no inclusions of lamellar morphology are detected. Therefore, these irradiation conditions cause the melting of all phases in the subsurface layer of silumin. The further high-speed cooling of the melted layer results in the formation of silicon inclusions with mainly globular morphology.

Metallographic studies have established that globules vary in a range of 5–20 μm. Globules are found along the alloy grain (30–50 μm) boundaries. A mottled structure is uncovered in the volume of grains (Fig. 3.9d), which might be considered a structure of cellular crystallization. Dimensions of mottled contrast elements are in a range of 300–500 nm.

Increasing the energy density of the electron beam to 20 J/cm^2, it is reported on the roughing of a forming structure (Fig. 3.11).

Additionally, SEM analysis was carried out to investigate the structure of the modified surface formed after being irradiated by an intense electron beam (15 J/cm^2, 150 μs, 3 impulses; and 20 J/cm2, 150 μs, 5 impulses).

Figure 3.12 demonstrates the research outcomes of the silumin surface irradiated by an electron beam (15 J/cm^2, 150 μs, 3 impulses). Analyzing the structure of silumin formed after being irradiated in the conditions above, conclusions deduced in the metallographic research (Fig. 3.8) were confirmed. For instance, electron-beam processing initiates two interrelated processes, i.e. globularization of silicon inclusions (Fig. 3.12a) and brittle defragmentation of silicon lamellae (Fig. 3.12b).

Once lamellae fracture, numerous micro-pores tend to emerge along the lamella-matrix boundary and micro-cracks in silicon lamellae appear (Fig. 3.12b). Apparently, a material with such structure has relatively low mechanical characteristics. Silicon lamellae with micro-pores and micro-cracks are stress risers, i.e. spots of micro-cracks originating. It will be demonstrated further that fatigue tests of samples modified by an electron beam (energy density 15 J/cm^2, 150 μs, 5 impulses) have pointed out low values of fatigue life.

To sum up, the irradiation of silumin surface by a high-intense pulsed electron beam in the melting mode of silicon inclusions causes the formation of micro-pores and micro-cracks in the material surface, weakening the material as a consequence. This fact seriously restricts the fatigue life of the material.

Figure 3.13 demonstrates typical images of the structure formed after being irradiated the surface of silumin by highly intense pulsed electron beam (energy density 20 J/cm^2, 150 μs, 5 impulses). As seen, the morphology of the surface structure

3.2 Structure and Phase Composition of Silumin Irradiated ... 85

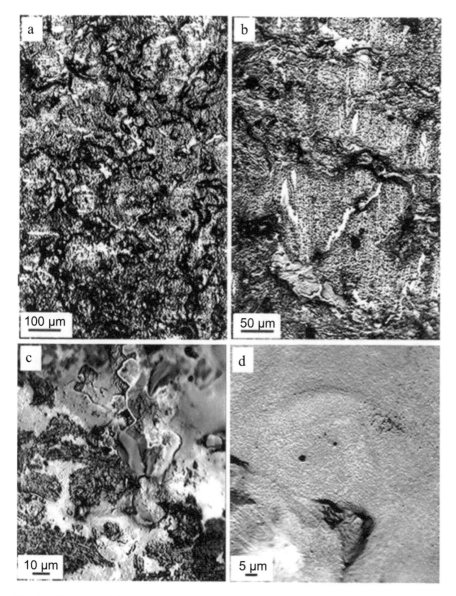

Fig. 3.9 Structure of the silumin surface irradiated by an intense pulsed electron beam (energy density 20 J/cm^2, 150 μs, 1 impulse)

differs greatly from the structure in the initial sample (Figs. 3.1 and 3.6) and a sample produced in conditions: energy density 15 J/cm^2, 150 μs, 3 impulses (Fig. 3.12). A homogenous grain-type (cellular) structure is formed on the irradiated surface (20 J/cm^2, 150 μs, 5 impulses); grains are in a range of 30–50 μm. The grains are surrounded by silicon layers, the maximal transversal dimensions of which are

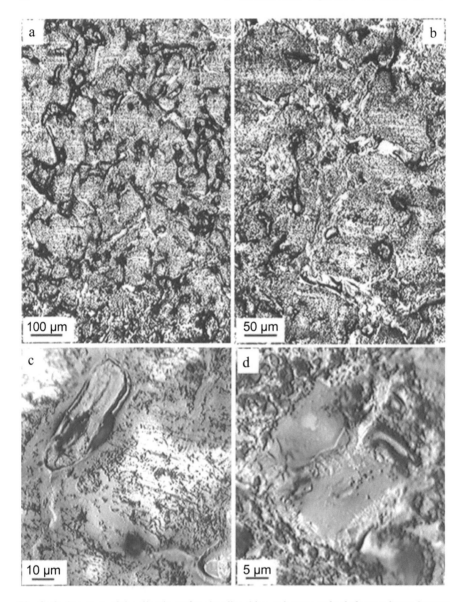

Fig. 3.10 Structure of the silumin surface irradiated by an intense pulsed electron beam (energy density 20 J/cm^2, 150 μs, 5 impulses)

20 μm (Fig. 3.13b). Stress risers (micro-pores, micro-cracks, micro-craters, brittle inclusions of intermetallic particles), which might cause the fracture of samples during further mechanical testing, are unfound on the irradiated surface.

Therefore, the research of as-cast silumin AK12 conducted by the methods of up to date materials study has demonstrated that this alloy is a multiphase material

3.2 Structure and Phase Composition of Silumin Irradiated … 87

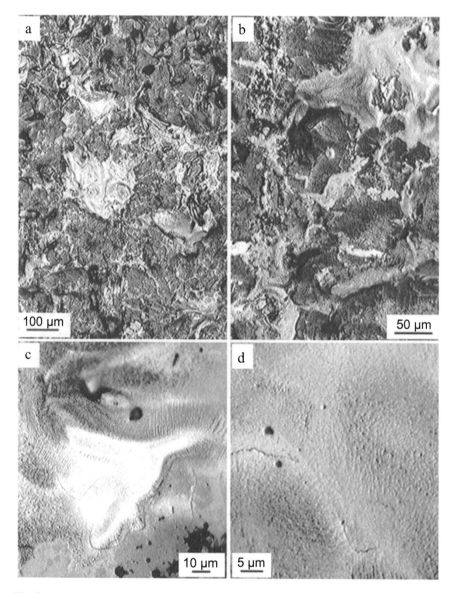

Fig. 3.11 Structure of the silumin surface irradiated by an intense pulsed electron beam (energy density 20 J/cm^2, 150 μs, 3 impulses)

composed of aluminum and silicon phases as well as intermetallic compounds Al–Si–Fe–Mn. Particles of silicon and intermetallic compounds possess a lamellar form or look like skeletons, as a consequence, being potential reasons for cracking during mechanical or thermal-mechanical testing of the material [4–12].

Fig. 3.12 Structure of the silumin surface irradiated by an electron beam (energy density 15 J/cm^2, 150 µs, 3 impulses)

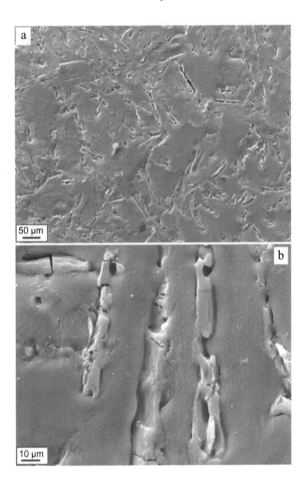

The study on the silumin surface irradiated by a highly intense pulsed electron beam has established parameters of an electron beam that cause a strong effect on structures being formed, i.e. their states may differ largely. That is, a surface layer with a great number of defects, and probably insufficient mechanical characteristics, and is formed in conditions initiating a first melting phase of silicon inclusions. In conditions, which support high-speed melting and further high-speed crystallization, a grain-type (cellular) structure is found in the surface layer, and globular silicon particles are detected on the grain boundaries, whereas nano-dimensional silicon layers are on cell boundaries. We expect that a material with such structure might exhibit better mechanical characteristics than as-cast material.

Fig. 3.13 Structure of the silumin surface irradiated by an electron beam (energy density 20 J/cm^2, 150 μs, 5 impulses)

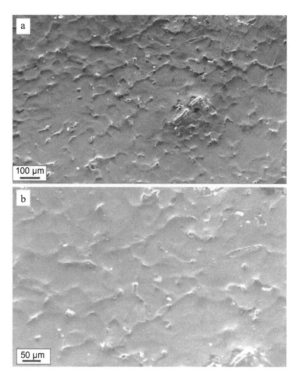

References

1. Poletika, I.M., et al.: Development of a new class of coatings by double electron-beam surfacing. Inorg. Mater. Appl. Res. **2**, 531–539 (2011)
2. Muratov, V.S., Morozova, E.A.: Titanium structure and property formation with chromium laser surface alloying. Met. Sci. Heat Treat. **61**, 340–343 (2019)
3. Yokobori, T.: Physics of Strength and Plasticity. MIT Press, Boston (1969)
4. Ivanov, Y.F., et al.: Fractography of the fatigue fracture surface of silumin irradiated by high-intensity pulsed electron beam. IOP Conf. Ser. Mater. Sci. Eng. **81**, 012011 (2015)
5. Ivanov, Y.F., Alsaraeva, K.V., Gromov, V.E., Popova, N.A., Konovalov, S.V.: Fatigue life of silumin treated with a high-intensity pulsed electron beam. J. Surf. Investig. X-ray, Synchrotron Neutron Tech. **9**, 1056–1059 (2015)
6. Gromov, V.E., Ivanov, Y.F., Glezer, A.M., Konovalov, S.V., Alsaraeva, K.V.: Structural evolution of silumin treated with a high-intensity pulse electron beam and subsequent fatigue loading up to failure. Bull. Russ. Acad. Sci. Phys. **79**, 1169–1172 (2015)
7. Ivanov, Y.F., Aksenova, K.V., Gromov, V.E., Konovalov, S.V., Petrikova, E.A.: An increase in fatigue service life of eutectic silumin by electron-beam treatment. Russ. J. Non-Ferrous Met. **57**, 236–242 (2016)
8. Konovalov, S.V., Alsaraeva, K.V., Gromov, V.E., Ivanov, Y.F.: Structure-phase states of silumin surface layer after electron beam and high cycle fatigue. J. Phys: Conf. Ser. **652**, 012028 (2015)
9. Konovalov, S., Alsaraeva, K., Gromov, V., Ivanov, Y., Semina, O.: Structure-phase states evolution in Al-Si alloy under electron-beam treatment and high-cycle fatigue. AIP Conf. Proc. **1683**, 020092 (2015)

10. Konovalov, S.V., Aksenova, K.V., Gromov, V.E., Ivanov, Y.F., Semina, O.A.: Fractography of fatigue fracture surface in silumin subjected to electron-beam processing. IOP Conf. Ser. Mater. Sci. Eng. **142**, 012080 (2016)
11. Konovalov, S.V., Aksenova, K., Gromov, V., Ivanov, Y.F., Semina, O.: The influence of electron beam treatment on Al-Si alloy structure destroyed at high-cycle fatigue. Key Eng. Mater. **675–676**, 655–659 (2016)
12. Ivanov, Y., Alsaraeva, K., Gromov, V., Konovalov, S., Semina, O.: Evolution of Al–19.4Si alloy surface structure after electron beam treatment and high cycle fatigue. Mater. Sci. Technol. **31**, 1523–1529 (2015)

Chapter 4
Fractography of Silumin Surface Fractured in High-Cycle Fatigue Tests

Chapter 1 has demonstrated that the fatigue failure of machine elements appears to be one of the most frequent reasons for the breakdown of equipment, mechanisms, machines, and facilities. This fact stems from peculiarities of the high-cycle fatigue. This is the initiation and development of a crack caused by a comparatively reduced stress, and the high sensitivity of fatigue endurance to various factors of design, technology, and maintenance. Another factor is that the fatigue resistance varies in an extensive range (dispersion of a lifespan) as compared with the static strength. Additionally, cracks are initiated and propagated randomly without any apparent displacement till an emergency situation. Therefore, the prevention from fatigue breakdowns of high-duty machine elements (the elongating of their service life) remains a vital issue, especially in industries, where an emergency may bring about fatal consequences.

Fatigue cracks tend to appear in the surface layer of a machine part, so a state of the surface layer is of extraordinary importance for the fatigue endurance of materials. The use of face hardening enhances considerably (by 2–3 times) the fatigue endurance due to the eliminating of machining micro-defects (scratch marks, scratches, roughness), the development of compressive residual stress in the surface layer of a machine part to be hardened, and the disintegrating of a matrix structure and second phase inclusions.

A significant increase in resource properties of surface layers in the material is possible if the results-oriented development of additional structure-phase state levels is realized in sub-micro and nano-scale ranges. An efficient method of such modification resulting in the elongation of a fatigue life is the processing of a material surface by an intense sub-millisecond pulsed electron beam, which changes the structure of a surface layer (up to tens of micrometers); as a consequence of this process, the surface is in a multimodal structure-phase state, whereas the structure-phase state of the main alloy volume is practically stable.

This chapter aims to analyze regularities and reveal mechanisms responsible for the fracturing of silumin AK12 in the process of high-cycle fatigue tests.

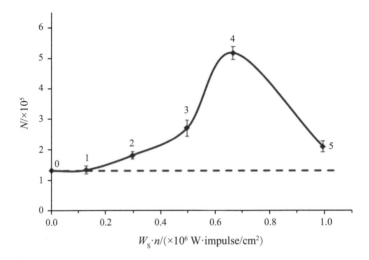

Fig. 4.1 The average number of cycles till the fracture <N> vs. the product of power density W_S and the number of impulses n generated by an electron beam. A dotted line indicates an average fatigue life of the initial (as-cast) material, reprinted from ref. [1] (Copyright 2014, with permission from Taylor & Francis)

Table 4.1 Irradiation conditions of silumin by pulsed electron beam and findings of high-cycle fatigue tests (conditions are numbered similarly to the graph in Fig. 4.1)

No.	$E_S/(J/cm^2)$	$\tau/\mu s$	n/impulse	$W_S \cdot n/(\times 10^6$ W·impulse/cm^2)	$<N>/\times 10^5$
1	20	150	1	0.13	1.32
2	15	150	3	0.30	1.80
3	25	150	3	0.51	2.70
4	20	150	5	0.67	5.17
5	10	50	5	1.00	2.09

The fatigue testing of samples in the initial state and irradiated by an intense pulsed electron beam was carried out using a specially developed laboratory unit according to the cyclic asymmetric cantilever bending (see Chapter 2). A front face of the sample, i.e. one under a crack imitating cut was processed by an electron beam. At least 5 samples were put to the test in each irradiation mode. To analyze a fatigue failure surface, researchers utilized SEM methods.

The fatigue tests have demonstrated a considerable dispersion of outcomes varying in a broad range (132,000–517,000 cycles till fracture). Having analyzed average values, a correlation of a fatigue life (a number of cycles till fracture) vs. a generalized parameter of an electron beam (a product of power density), $W_S \cdot n$ was obtained (Fig. 4.1, Table 4.1). To get an insight into the fatigue failure mechanism, first untreated silumin is considered.

4.1 Fractography of a Fatigue Failure Surface in as-Cast Silumin

As argued, a feature of as-cast silumin AK12 is principally lamellar silicon crystallites in large quantities (Fig. 4.2).

Lamellae are arranged chaotically or they decorate alloy grain boundaries. Lamellae of silumin under consideration vary in a range from several to tens of micrometers. The fatigue failure is thought to represent a process developing locally in a material. When a material is in a critical state, it fractures. Two characteristic zones are found on a fracture surface—a zone of fatigue crack growth and a rupture area separated by an area of the rapid crack growth [2, 3]. The processes of deformation typical for the fatigue testing of the material tend to develop extensively in a zone of fatigue crack growth and far less intensively in a rupture area.

Figure 4.3 shows a characteristic fracture surface of as-cast silumin AK12 fractured as a result of 130,000 cycles. The research has demonstrated that a zone of fatigue crack growth in a sample fractured as a result of 130,000 cycles approximates to 1.8 mm.

The height of the zone of fatigue crack growth is considered to be a critical crack length. Further, this rupture characteristic of as-cast silumin AK12 will be compared with that of silumin irradiated by a pulsed electron beam in various conditions. A material safety factor might be determined approximately according to the relation between a fatigue failure zone and a rupture area: the lower this value, the lower the safety factor for similar loads during fatigue testing. This factor for a rupture given in Fig. 4.3 is as high as 0.53.

The significant but quite localized deformation is registered in each loading cycle at a crack top [2, 3]. A rupture front of the material tends to the thick branching due to the polycrystalline structure (grain structure of a solid aluminum-based solution

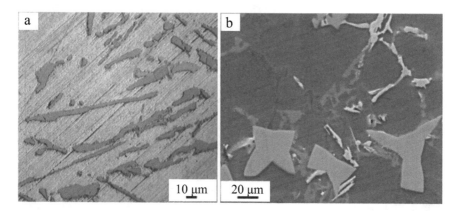

Fig. 4.2 Structure of silumin before fatigue tests (in initial state): (**a**) optical microscopy; (**b**) SEM

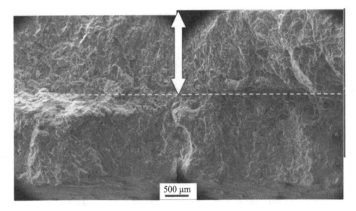

Fig. 4.3 Fractography of failure surface in as-cast silumin AK12. Arrows indicate a zone of fatigue crack growth

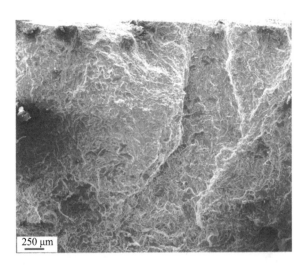

Fig. 4.4 Structure of failure surface in as-cast silumin sample

and relatively coarse silicon lamellae in large quantities). It is reported on the development of a vast number of microscopically detected parallel fracture signs (shown in Fig. 4.4).

Fatigue striations are thought to be a characteristic of a fatigue failure zone in the material. A characteristic silumin failure surface with fatigue striations is demonstrated in Fig. 4.5.

Fatigue striations are generally understood to mean strips of successively located grooves and peaks or strips with steps surrounded by these grooves to be located parallel to a crack front (Fig. 4.5c, d). Each loading cycle moves forward a crack (rupture) to a certain distance. At this place, a series of strips are produced on a fracture surface. Therefore, strips may be considered generally as a trace of a crack moving one step per loading cycle. Schmidt-Thomas and Klingele suggested calling

4.1 Fractography of a Fatigue Failure Surface in as-Cast Silumin

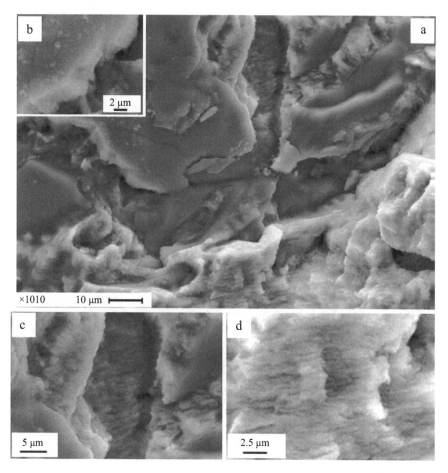

Fig. 4.5 Structure of failure surface in as-cast silumin sample

these strips fatigue striations. They are perpendicular or almost perpendicular to the direction of a propagating crack. Striations might be continuous and regular (a characteristic of aluminum alloys); distances between them tend to get shorter if the stress drops and the propagation of a crack slows down. They might be broken and irregular like in fracture surfaces of steel.

The gap between striations correlates with the resistance of a material to the fatigue crack growth; all other experimental conditions of fatigue loading kept constant: the smaller the spacing between strips, the higher the resistance to the crack propagation a material has. Our research has revealed that the average distance between fatigue striations in fractured samples of as-cast silumin AK12 is 0.75 μm.

In most cases, a fracture surface is distinguished by a complex structure. The mechanism of fatigue failure in two-phase materials like silumin under study is

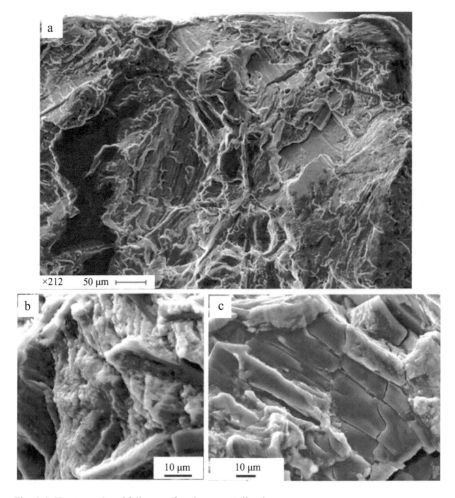

Fig. 4.6 Fractography of failure surface in as-cast silumin

complex. The data in Fig. 4.6 show grooves of ductile failure and facets of quasi-spalling. The grooves, being a core structural element of a fracture surface, develop because micro-pores furthering aluminum grain fracture are cut (Fig. 4.6b). Silicon lamellae fracture according to a rupture mechanism (Fig. 4.6c).

4.2 Fractography of Fatigue Failure Surface in Silumin Irradiated by an Intense Pulsed Electron Beam

The structure of a surface layer might be seriously modified owing to the irradiation by an intense pulsed electron beam (see Sect. 3.2). Furthermore, outcomes of the modifying depend on the density of energy and a number of impulses a beam of

4.2 Fractography of Fatigue Failure Surface in Silumin Irradiated ...

electrons produces. It is completely expectably that mechanical characteristics of the modified silumin, e.g. the fatigue endurance, will also be connected with the previously mentioned parameters of an electron beam.

To research into the structure of a silumin fracture surface modified by an electron beam, samples with a minimal ($N_1 = 180{,}000$) and maximal ($N_2 = 517{,}000$) number of cycles till the fracture were selected. The value N_1 was determined for samples irradiated by an electron beam in conditions 15 J/cm^2, 150 μs, 3 impulses; the value N_2—20 J/cm^2, 150 μs, 5 impulses, respectively. The research data on the untreated silumin fracture surface (in the initial state) are presented in Sect. 4.1.

4.2.1 Analysis of a Fracture Surface in Silumin Samples Modified by an Electron Beam with a Minimal Fatigue Life

As stated above, a fatigue life of a material is in a direct connection with the structure of a surface layer in the sample being tested. Figure 4.7 provides the research data on the silumin surface structure irradiated by an electron beam in conditions 15 J/cm^2, 150 μs, 3 impulses, which displayed the shortest fatigue life during fatigue testing.

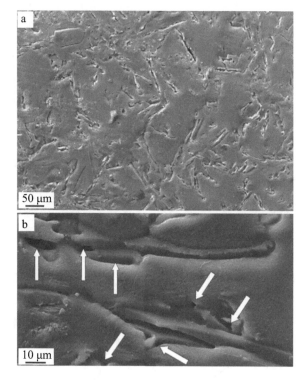

Fig. 4.7 Structure of a silumin surface sample processed by an electron beam in conditions 15 J/cm^2, 150 μs, 3 impulses; in a before tests state. Arrows (**b**) indicate micro-pores developed in the high-speed crystallization of a molten surface layer

Analyzing a structure formed under the irradiation of silumin in conditions above, a conclusion has been made that electron-beam processing leads only to the partial melting of excessive silicon lamellae (Fig. 4.7a). The melting of lamellae facilitates the development of micro-pores in large quantities, found along the boundary lamella/matrix and micro-cracks in silicon lamellae (Fig. 4.7b). Silicon lamellae remain stress risers. Fatigue tests cause the fracture of lamellae (Fig. 4.8b) and the initiation of long micro-cracks (Fig. 4.8a).

Therefore, the irradiation of a silumin surface by a high-intense pulsed electron beam in the melting mode of silicon lamellae gives rise to the generation of micro-pores and micro-cracks in the surface layer, weakening the material, as a consequence. The latter represents a key factor to shorten the fatigue endurance of the material (see Fig. 4.1).

During fatigue testing, cracks tend to be initiated on the surface of a sample or in the subsurface layer. Having investigated a surface layer structure in silumin irradiated in conditions 15 J/cm^2, 150 µs, 3 impulses and demonstrating considerably short fatigue endurance ($N_2 = 180,000$), a reason for the material fracture was found.

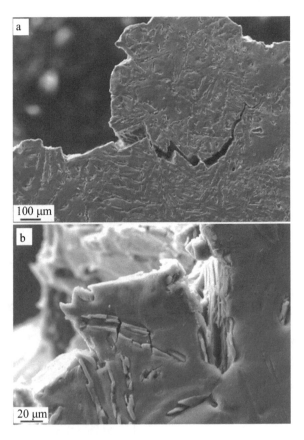

Fig. 4.8 Structure of a silumin surface sample processed by an electron beam in conditions 15 J/cm^2, 150 µs, 3 impulses and fractured as a result of 180,000 cycles

out. As expected, the risers of critical stress are large silicon lamellae in the surface (Fig. 4.8) and the subsurface layer (Fig. 4.9) of a sample.

Figure 4.10 provides a characteristic view of a fracture surface in silumin AK12 samples irradiated in conditions 15 J/cm^2, 150 μs, 3 impulses, 0.3 s^{-1} and fractured as a result of 180,000 cycles. A thickness of the zone of fatigue crack growth in the sample fractured after 180,000 cycles was determined to be ≈0.75 mm.

The prior research has suggested a ratio of a fatigue zone to a rupture area might give an approximate safety factor of the material under study [4]: the lower the ratio, the worse a safety factor, and a load during fatigue testing kept stable. This parameter for a rupture in Fig. 4.10 is assessed to be 0.24.

The views of a silumin fracture surface in Figs 4.10 and 4.11 demonstrate many parallel fracture signs detected by a microscope, which indicate the thick branching in the rupture front in the material.

This structure of a fracture surface is typical for polycrystalline materials. A grain structure in an aluminum-based solid solution and many relatively large silicon lamellae cause the thick branching of the material fracture front, which creates the complex fracture topography.

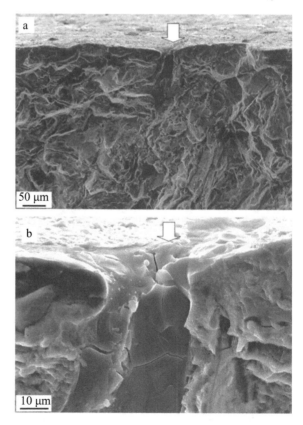

Fig. 4.9 Fractography of a silumin fatigue failure surface processed by an electron beam in conditions 15 J/cm^2, 150 μs, 3 impulses and fractured as a result of 180,000 cycles. Arrows indicate silicon lamellae

Fig. 4.10 Fractography of a silumin fatigue failure surface processed by an electron beam in conditions 15 J/cm^2, 150 μs, 3 impulses and fractured as a result of 180,000 cycles. Arrows indicate a zone of fatigue crack growth

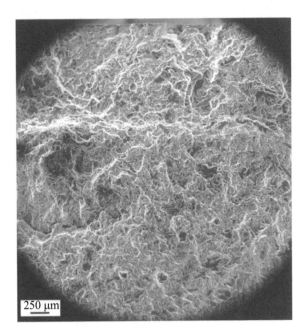

Fig. 4.11 Structure of a silumin fracture surface processed by an electron beam in conditions 15 J/cm^2, 150 μs, 3 impulses and fractured as a result of 180,000 cycles

Researchers have argued that fatigue striations are important characteristics in a fatigue zone of a material fracture [4–6]. Figure 4.12 provides a characteristic view of a fracture surface in silumin AK12 irradiated in conditions 15 J/cm^2, 150 μs, 3 impulses and fractured as a result of 180,000 cycles and with fatigue striations.

We have already mentioned that the space between striations is connected with the resistance of a material to the fatigue crack growth, all other experimental conditions of fatigue testing kept constant: the narrower the spacing between strips, the higher the resistance to the crack propagation a material has. An average distance between fatigue striations in fractured samples of silumin AK12 irradiated in conditions 15 J/cm^2, 150 μs, 3 impulses was estimated to be 0.95 μm.

Fig. 4.12 Fatigue striations in silumin developed due to the fatigue failure as a result of 180,000 cycles

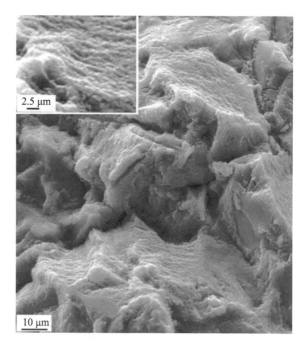

The data in Fig. 4.13 (a characteristic electro-microscopic view of a fracture surface) show grooves of ductile failure and facets of quasi-spalling. The grooves, being a key structural element of a fracture surface, develop because micro-pores furthering aluminum grain fracture are cut. Silicon lamellae fracture according to a rupture mechanism.

4.2.2 Analysis of a Fracture Surface in Silumin Samples Modified by an Electron Beam with a Maximal Fatigue Life

Figure 4.14 a, b offer characteristic views of a structure developing when silumin is irradiated by a high-intense pulsed electron beam in conditions 20 J/cm^2, 150 μs, 5 impulses, which has demonstrated the longest fatigue life if put to fatigue tests (517,000 cycles). The figure clearly shows there is a wide gap between the morphology of the surface layer structure in the untreated sample (Fig. 3.1) and the sample irradiated in conditions 15 J/cm^2, 150 μs, 3 impulses (Fig. 3.12). A homogenous grain (cellular) structure is formed on the irradiation surface with grains varying in a range of 30–50 μm. Between grains there are silicon layers detected with cross sizes below 20 μm (Fig. 4.14b). The research has found no stress risers on the edge of a rupture, which might cause a fracture of a sample (Fig. 4.14c). Cracks found

Fig. 4.13 Fractography of a silumin fatigue failure surface processed by an electron beam in conditions 15 J/cm^2, 150 μs, 3 impulses and fractured as a result of 180,000 cycles. Arrows indicate the surface of silumin irradiated by a high-intense pulsed electron beam

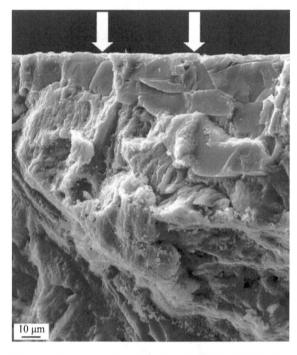

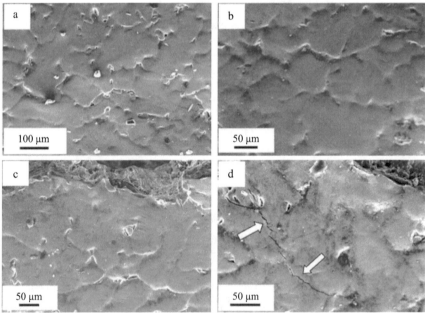

Fig. 4.14 Structure of silumin surface irradiated by an electron beam in conditions 20 J/cm^2, 50 μs, 5 impulses: (**a**)(**b**) in a before fatigue tests state; (**c**)(**d**) tested surfaces (517,000 cycles). Arrows (**d**) indicate a micro-crack developed in the process of fatigue tests

parallel to the fracture surface are at some distance from it (Fig. 4.14d). It likely indicates that a stress riser, which caused fracture of a sample, was beneath the surface, apparently, on the boundary between liquid and solid phases.

To sum up, the research into the silumin surface irradiated by a high-intense pulsed electron beam has pointed out that a fatigue life of treated silumin can be approximately 3.5 times extended than that of as-cast silumin (Fig. 4.1) due to the high-speed melting and the subsequent crystallization of the surface layer, which results in the development of a cellular structure and long layers or globular inclusions of silicon on the boundaries of cells.

In previous paragraph, it has been suggested that the fatigue failure represents a temporary process taking place in local areas of the material. A certain critical state leads to the fracture of a whole sample. Two characteristic zones are found on a fracture surface—a zone of fatigue crack growth and a rupture area separated by an area of the fast crack growth. The deformation typical for the fatigue testing of the material tends to develop extensively in a zone of fatigue crack growth and far less intensively in a rupture area. Figure 4.15b provides a characteristic view of a fracture surface in silumin sample fractured after 517,000 cycles. The thickness of the zone of fatigue crack growth is estimated to be 3.5 mm in the sample fractured as a result of 517,000 cycles (Fig. 4.15b).

Comparing these data with the prior results (a zone of fatigue crack growth in a sample fractured after 180,000 cycles is ≈0.75 mm (Figs. 4.10 and 4.15a), a conclusion may be drawn that this characteristic of a silumin fracture surface correlates with a number of cycles till fracture, i.e. it is associated with irradiation conditions of the material by an electron beam. A height, which a zone of fatigue crack growth

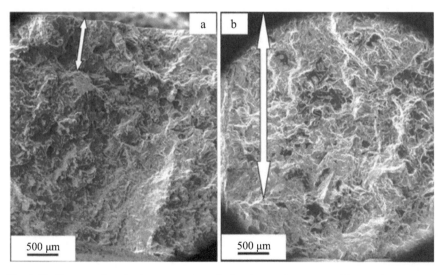

Fig. 4.15 Fractography of a fracture surface of silumin samples processed by an electron beam and fractured as a result of $N_1 = 180,000$ cycles (**a**) and $N_2 = 517,000$ cycles (**b**). Arrows indicate a zone of fatigue crack growth

has, is considered to be similar to a critical length of a crack. Therefore, an optimally selected electron-beam processing mode of the silumin surface might result in a more than 3 times increase of a critical crack length, extending the efficiency resource of the material, as a consequence. It has been reported on similar results when high-cycle loading of ferritic and pearlitic steel 60ГС2 exposed to electro-pulse processing at a certain stage [7, 8].

As brought to the light in the prior research, a ratio of a fatigue zone to a rupture area might give an approximate safety factor of the material under study: the lower the ratio, the worse the safety factor, and a load during fatigue testing keeps stable. Fractography data on the material under study (Figs. 4.10 and 4.15) have provided an evidence that this factor ranges from 0.24 (15 J/cm^2, 150 μs, 3 impulses) to 0.86 (20 J/cm^2, 150 μs, 5 impulses). Therefore, the irradiation of silumin in appropriate conditions raises significantly a safety factor of the material.

During fatigue testing, cracks tend to be initiated on the surface of a sample or in the subsurface layer. The study on the surface layer structure in silumin irradiated in conditions 15 J/cm^2, 150 μs, 3 impulses and demonstrating considerably short fatigue endurance ($N_1 = 132{,}000$ and $N_2 = 180{,}000$) has revealed a reason for the material fracture. As shown above (Fig. 4.9), the risers of critical stress are large silicon lamellae in the surface and the subsurface layer (Fig. 4.9) of a sample. When the silumin surface is irradiated by an electron beam in conditions 20 J/cm^2, 150 μs, 5 impulses, the melting of a surface layer with a thickness of 20 μm and above is registered (Fig. 4.16a). High-speed crystallization conduces to the development of structure with crystallites varying in a range of 250–100 nm (Fig. 4.16b). Apparently, this sub-micro and nano-dimensional structure facilitates a multiple increase of the silumin fatigue endurance.

The polycrystalline structure (grain structure of a solid aluminum-based solution and relatively coarse silicon lamellae in large quantities) causes the thick branching of a rupture front in the material. It is reported on the development of a great number of microscopically detected parallel fracture signs; it is typical for a sample with a maximal number of cycles till fracture (Fig. 4.17).

In previous paragraph, fatigue striations were argued to represent a critical characteristic of a fracture zone in the material. Figure 4.18 presents a typical view of a silumin fracture surface with fatigue striations.

The space between striations is attributed to the resistance of a material to the fatigue crack growth: the narrower the spacing between strips, the higher the resistance to the crack propagation a material has. The study has pointed out that an average distance between fatigue striations in samples of silumin fractured after 180,000 cycles is 0.95 μm, and in a sample fractured after 517,000 cycles is 0.28 μm, respectively. Therefore, a crack step per loading cycle in a silumin sample processed in optimal conditions is 3.5 times smaller, so this sample has far higher resistance to the propagation of a fatigue crack.

The data on a rupture structure (Fig. 4.19) show grooves of ductile failure and facets of quasi-spalling. The grooves, representing a main structural element of a fracture surface, develop because micro-pores furthering aluminum grain fracture are cut (Fig. 4.19b). Silicon lamellae fracture according to a spalling mechanism.

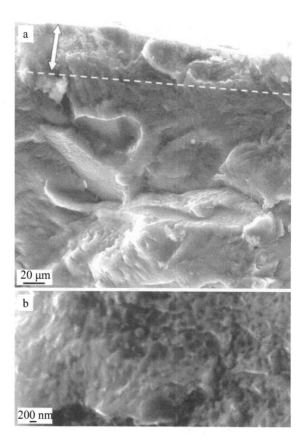

Fig. 4.16 Fractography of a silumin fracture surface as a result of 517,000 cycles. Arrows (**a**) indicate a layer to crystallize from a state molten by an electron beam

To conclude, this chapter presents the data on the evolution of a fracture surface in silumin samples modified by a high-intense electron beam. The mode of irradiation enhancing the fatigue endurance of the material by approximately 3.5 times was determined. The structures of irradiation and fatigue failure surfaces in as-cast (untreated) silumin and as-cast silumin modified by an intense pulsed electron beam were analyzed. The principal reason for the enhancement of the silumin fatigue endurance was found out to be dispersion and a quasi-homogenous pattern of silicon crystals in the modified layer. An assumption was made that critical stress risers to form in the subsurface layer on the boundary of liquid and solid phases under optimal conditions of irradiation [1, 9–17].

The study demonstrates that the modification of silicon lamellae furthers the development of micro-pores in large quantities in the mode of partial melting, which are detected along the boundary lamella/matrix and micro-cracks in silicon lamellae. In the mode of stable melting (a molten layer ranges up to 20 μm), a multimodal structure develops (dimensions of grains in a range of 30–50 μm, dimensions of silicon particles on their boundaries up to 10 μm, and a sub-grain structure in a form of crystallization cells (100–250 nm)). A fatigue life of silumin is shown to extend due

Fig. 4.17 Fractography of a silumin fracture surface processed by an electron beam in conditions 20 J/cm^2, 150 µs, 5 impulses and fractured as a result of 517,000 cycles

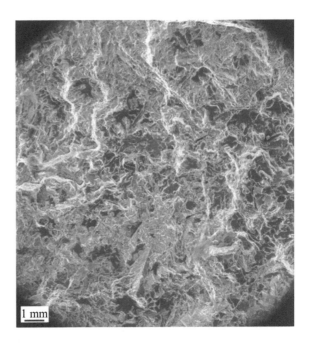

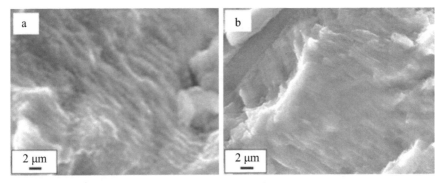

Fig. 4.18 Fatigue striations developing in silumin samples as a result of fatigue fracture: (**a**) a sample fractured as a result of 130,000 cycles; (**b**) a sample fractured as a result of 517,000 cycles

to a considerable increase in a critical length of a crack and a safety factor, as well as the more compact distance between fatigue striations (a crack step per loading cycle) and the development of a multimodal sub-micro and nano-dimensional multiphase structure.

Fig. 4.19 Structure of a silumin fracture surface processed by an electron beam in conditions 20 J/cm², 150 μs, 5 impulses and fractured as a result of 517,000 cycles

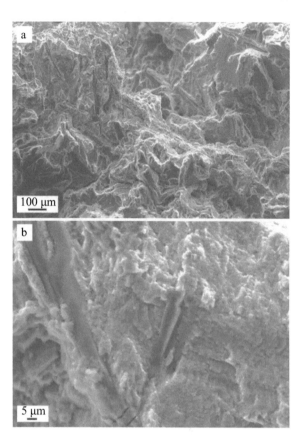

References

1. Ivanov, Y., Alsaraeva, K., Gromov, V., Konovalov, S., Semina, O.: Evolution of Al–19.4Si alloy surface structure after electron beam treatment and high cycle fatigue. Mater. Sci. Technol. **31**, 1523–1529 (2015)
2. Bagmutov, V.P., Parshev, S.N.: Integrated approach to the electromechanical formation of a structurally inhomogeneous surface layer on steel parts. Steel Transl. **34**, 66–69 (2004)
3. Zhukeshov, A.M., Gabdullina, A.T., Amrenova, A.U., Ibraimova, S.A.: Hardening of structural steel by pulsed plasma treatment. J. Nano-Electron. Phys. **6**, 03066 (2014)
4. Cui, W.: A state-of-the-art review on fatigue life prediction methods for metal structures. J. Mar. Sci. Technol. **7**, 43–56 (2002)
5. Stephens, R.I., Fatemi, A., Stephens, R.R., Fuchs, H.O.: Metal Fatigue in Engineering. Wiley Interscience, Hoboken, NJ (2000)
6. Song, L.X., Zhang, K.M., Zou, J.X., Yan, P.: Surface modifications of a hyperperitectic Zn-10 wt% Cu alloy by pulsed electron beam treatment. Surf. Coat.S Technol. **388**, 125530 (2020)
7. Sosnin, O.V. et al.: Structural and phase transformations in austenitic steel. Tyazheloe Mashinostr. **2**, 25–29 (2003)
8. Sosnin, O.V. et al.: Evolution of ferrite-perlite structure under the action of electric current pulses. Fiz. i Khimiya Obrab. Mater. **4**, 63–69 (2003)
9. Ivanov, Y.F. et al.: On the fatigue strength of grade 20Cr13 hardened steel modified by an electron beam. J. Surf. Investig. X-ray, Synchrotron Neutron Tech. **7**, 90–93 (2013)

10. Ivanov, Y.F., et al.: Fractography of the fatigue fracture surface of silumin irradiated by high-intensity pulsed electron beam. IOP Conf. Ser. Mater. Sci. Eng. **81**, 012011 (2015)
11. Ivanov, Y.F., Alsaraeva, K.V., Gromov, V.E., Popova, N.A., Konovalov, S.V.: Fatigue life of silumin treated with a high-intensity pulsed electron beam. J. Surf. Investig. X-ray, Synchrotron Neutron Tech. **9**, 1056–1059 (2015)
12. Gromov, V.E., Ivanov, Y.F., Glezer, A.M., Konovalov, S.V., Alsaraeva, K.V.: Structural evolution of silumin treated with a high-intensity pulse electron beam and subsequent fatigue loading up to failure. Bull. Russ. Acad. Sci. Phys. **79**, 1169–1172 (2015)
13. Ivanov, Y.F., Aksenova, K.V., Gromov, V.E., Konovalov, S.V., Petrikova, E.A.: An increase in fatigue service life of eutectic silumin by electron-beam treatment. Russ. J. Non-Ferrous Met. **57**, 236–242 (2016)
14. Konovalov, S.V., Alsaraeva, K.V., Gromov, V.E., Ivanov, Y.F.: Structure-phase states of silumin surface layer after electron beam and high cycle fatigue. J. Phys: Conf. Ser. **652**, 012028 (2015)
15. Konovalov, S., Alsaraeva, K., Gromov, V., Ivanov, Y., Semina, O.: Structure-phase states evolution in Al-Si alloy under electron-beam treatment and high-cycle fatigue. In: AIP Conference Proceedings, vol. 1683, 020092 (2015)
16. Konovalov, S.V., Aksenova, K.V., Gromov, V.E., Ivanov, Y.F., Semina, O.A.: Fractography of fatigue fracture surface in silumin subjected to electron-beam processing. IOP Conf. Ser. Mater. Sci. Eng. **142**, 012080 (2016)
17. Konovalov, S.V., Aksenova, K., Gromov, V., Ivanov, Y.F., Semina, O.: The influence of electron beam treatment on Al-Si alloy structure destroyed at high-cycle fatigue. Key Eng. Mater. **675–676**, 655–659 (2016)

Chapter 5
Degradation of Silumin Structure and Properties in High-Cycle Fatigue Tests

To date, electron-beam processing is seen as a unique instrument of high efficiency for the research into the physical nature of structure-phase states formation and the target modification of silumin structure and properties to improve its operational characteristics.

As discussed in Chapter 1, the processing of a surface in stainless steels by a high-intensity pulsed electron beam results in a considerable (more than 3.5 times) increase of their fatigue life. Apparently, a deformation impact occurring in fatigue tests changes structure-phase states, strength, and tribological characteristics of the surface. The purpose of this chapter is to examine the behavior of structure-phase states and surface properties of silumin preliminary processed by electron beams in the process of fatigue loading.

5.1 Degradation of Silumin Properties Irradiated by an Electron Beam in High-Cycle Fatigue Tests

In the prior research, fatigue tests were carried out according to the cyclic asymmetric cantilever bending (see Chapter 2). To irradiate the surface of samples prepared for fatigue tests, a unit "SOLO" was used. Structure-phase states and silumin surface properties were examined on samples with minimal (irradiation conditions 20 J/cm^2, 150 μs, 1 impulse—132,000 cycles) and maximal (irradiation conditions 20 J/cm^2, 150 μs, 5 impulses—517,000 cycles) fatigue lives.

The data on wear resistance of silumin samples obtained when testing are presented in Fig. 5.1 (dark bars). Samples exposed to the fatigue testing for 132,000 cycles (Fig. 5.1, mode 1) displayed a minimal wear rate. As a result of 517,000 cycles in fatigue tests, a wear rate of the material tends to rise (Fig. 5.1, mode 2) remaining, however, less than the parameter of the input (untreated before fatigue tests) material (Fig. 5.1, mode 0). A friction coefficient behaves practically similarly (Fig. 5.1, light bars).

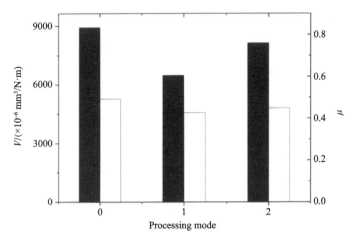

Fig. 5.1 Wear rate V (dark bars) and friction coefficient values μ (light bars) in various processing modes of silumin: 0—untreated silumin (initial state); 1—irradiation (20 J/cm^2, 150 µs, 1 impulse) and further fatigue tests (132,000 cycles); 2—irradiation (20 J/cm^2, 150 µs, 5 impulses) and further fatigue tests (517,000 cycles)

Mechanical properties of samples subjected to the fatigue testing were examined via assessing their micro- and nano-hardness. The microhardness was estimated employing the Vickers method by an appliance HVS-1000A (indenter load $P = 100$ mN). Microhardness profiles of modified samples are given in Fig. 5.2.

The data collected when measuring the microhardness of cross-sectional microsections (Fig. 5.2) allow a conclusion that a thickness of the hardened layer is ≈ 30 µm and its microhardness is 2–5 times higher than in the base material depending on the mode of electron-beam processing. The microhardness is related to the energy density of an electron beam. For instance, an increase in the energy density of the electron beam from 15 J/cm^2 to 25 J/cm^2 (Fig. 5.2, curve 2 and curve 3, respectively) enhances the microhardness. A low value of microhardness in the irradiation mode 10 J/cm^2, 150 µs, 5 impulses (Fig. 5.2, curve 5) and a maximal total energy density $W_S \cdot n = 1 \times 10^6$ W·impulses/cm^2 (Table 4.1) might be caused by the intensive remelting of the surface modified layer. Varying a number of fatigue loading cycles from 132,000 to 270,000, an approximately 1.6 times increase in silumin microhardness is achieved, however, the further loading to 517,000 cycles results in a substantial drop of microhardness.

The nano-hardness of the irradiated surface was determined for an indenter load ranging from 5 mN to 300 mN by an appliance NANO Hardness Tester NHT-S-AX-000X (the method of nano-indenting). Values of nano-hardness and Young modulus were estimated based on Oliver-Pharr method using the software of nano-hardness tester. Importantly, a hardness measurement error caused by distorted edges of an imprint may amount to 10% [1]. Taking into consideration the distortion of an imprint form, a result with an error of 4.7% may be acquired when estimating a hardness

5.1 Degradation of Silumin Properties Irradiated …   111

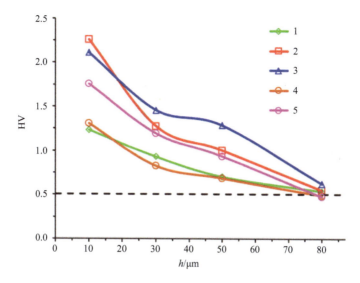

Fig. 5.2 Microhardness profiles of silumin samples irradiated by an intensive electron beam and loaded till a fracture in high-cycle fatigue tests. Numbers indicate modes of electron-beam processing: 1—20 J/cm^2, 150 μs, 1 impulse; 2—15 J/cm^2, 150 μs, 3 impulses; 3—25 J/cm^2, 150 μs, 3 impulses; 4—20 J/cm^2, 150 μs, 5 impulses; 5—10 J/cm2, 150 μs, 5 impulses. Indenter load 100 mN. A dotted line indicates the microhardness of as-cast silumin

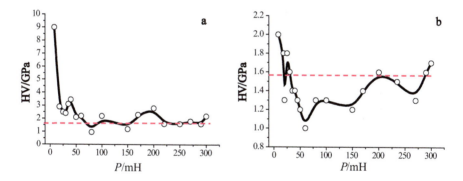

Fig. 5.3 Nano-hardness of the silumin surface layer processed by a high-intensity pulsed electron beam and tested till a fracture vs. indenter load: (**a**) 132,000 cycles; (**b**) 517,000 cycles. A dotted line indicates the hardness of as-cast silumin

value as compared with a value calculated according to a standard formula without the distortion of an imprint form.

A function of average nano-hardness vs. indenter load is plotted based on the findings of the test procedure described above (Fig. 5.3). As seen, if the number of fatigue loading cycles is approximately 4 times higher, the hardness of the surface layer is about 4.5 times lower. In the previous paragraph, it was shown that the nano-indenting allows the estimation of a material Young modulus. It has been established that a

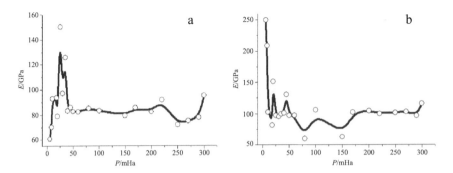

Fig. 5.4 Young modulus of the silumin surface layer processed by a high-intensity pulsed electron beam and tested till a fracture vs. indenter load: (**a**) 132,000 cycles; (**b**) 517,000 cycles

value of silumin Young modulus changes similarly to the nano-hardness (Fig. 5.4): a raise of fatigue testing cycles brings about a multifold decrease (3–4 times) in Young modulus of the material surface layer.

Comparing the material hardness at various indenter loads (microhardness (Fig. 5.2) and nano-hardness (Fig. 5.3) measurement), a vast gap (1.5–3 times) in quantitative data is found in spite of their quality compatibility. In works [2, 3], researchers discuss possible reasons for the quite high hardness of the material typical for the nano-indenting. The material is assumed to demonstrate the elastic behavior up to theoretically calculated stresses because of considerable deformation constraints. This behavior of the material (a tendency to demonstrate abnormally high microhardness) is referred to as dislocation starvation. The matter of this assumption is that a material imperfection degree under the imprint is considerably low due to its microscopic sizes (tens of nanometers) during nano-contacting; therefore, a real material tends to behave like an ideal one with values of microhardness similar to theoretically possible. It is recommended that the further research should be undertaken with the focus on a series of separate phenomena, types of atomic defects and their participants, importance, and contribution of these defects to the plastic deformation (the material deformation under the indenter).

Undoubtedly, tribological and strength characteristics of the material change owing to the transformations of the phase composition and defect substructure of the silumin surface layer during fatigue testing.

5.2 Evolution of Defect Substructure and Phase State of Silumin Irradiated by an Intensive Pulsed Electron Beam During Fatigue Testing

In this section, an evolution analysis of the silumin defect substructure irradiated by a high-intensity electron beam was carried out during fatigue testing. The fracture surface was examined by the methods of SEM; the defect structure of fractured

5.2 Evolution of Defect Substructure and Phase State of Silumin Irradiated ...

samples was analyzed using electron-diffraction microscopy of thin foils. Foils were prepared via ion thinning from plates cut parallel to the rupture surface at a minimal possible depth from the fracture surface.

The prior extensive discussion of as-cast silumin AK12 has shown it contains many silicon crystallites, principally in a form of lamellae (Fig. 5.5a). Lamellae are arranged chaotically or develop colonies. Silicon lamellae in a micro-section plane vary from several to tens of micrometers.

SEM of the modified surface has revealed the irradiation of silumin by an electron beam in conditions 20 J/cm^2, 150 μs, 1 impulse initiates the partial melting of silicon crystals in the surface layer; additionally, micro-pores develop around lamellae (Fig. 5.5b). In fatigue tests, silicon lamellae are stress risers and lead to an early fracture of the material (Fig. 5.5d, arrows indicate a crack developed along a silumin lamella).

The irradiation of silumin by an electron beam in conditions 20 J/cm^2, 150 μs, 5 impulses facilitates the globularization of silicon inclusions. In the surface layer a structure develops, where silicon in a form of globular inclusions is basically detected on grain boundaries (Fig. 5.5c). This fact appears to be a key factor extending significantly a fatigue life of the material.

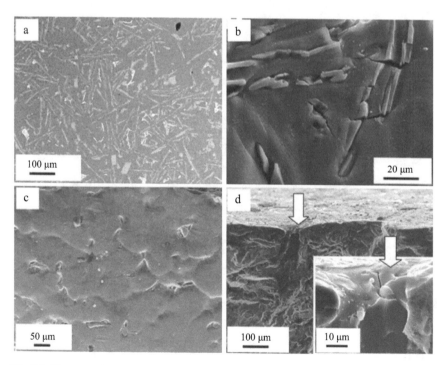

Fig. 5.5 Electron-microscopic view of as-cast silumin structure (**a**), irradiated by an electron beam in conditions 20 J/cm^2, 150 μs, 1 impulse (**b, d**) and 20 J/cm^2, 150 μs, 5 impulses (**c**); arrows in (**d**) indicate a micro-crack developed on the irradiated surface during fatigue testing

The melting of the silumin surface layer by an intensive pulsed electron beam and the further high-speed crystallization (irrespectively to irradiation modes) result in the formation of the columnar eutectics detected in SEM (Fig. 5.6a) and TEM (Fig. 5.6b–d) studies.

As expected, a columnar structure is composed of two phases and comprises layers of a solid aluminum-based solution separated by silicon layers (Fig. 5.6b). In most cases, a structure of high-speed crystallization has a cellular texture (Fig. 5.6c). Cells are 450 nm on average, cross sizes of silicon layers are 80 nm. In the paragraph above, it was suggested that the irradiation of silumin by an electron beam in conditions 20 J/cm^2, 150 µs, 1 impulse initiates the formation of micro-pores along the surface of silicon lamellae. An analysis of cross ruptures has indicated micro-pores in a subsurface with a columnar structure (Fig. 5.6a, pores are shown by arrows). The TEM-based research has pointed out centers of eutectics crystallization to represent a

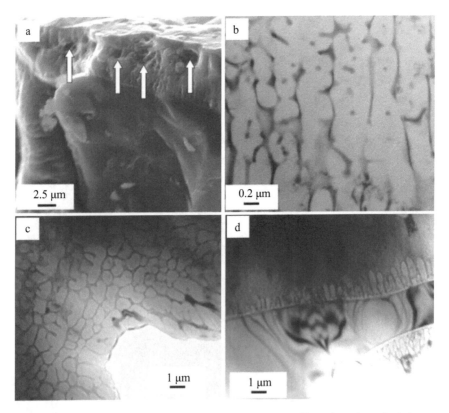

Fig. 5.6 Electron-microscopic view of silumin structure processed by an intensive pulsed electron beam in conditions 20 J/cm^2, 150 µs, 1 impulse and fractured as a result of 132,000 cycles of fatigue loading: (**a**) SEM; (**b**)–(**d**) TEM; arrows in (**a**) indicate pores developing in the surface layer of silumin irradiated by an electron beam

5.2 Evolution of Defect Substructure and Phase State of Silumin Irradiated ...

free surface in a sample as well as inclusions of intermetallic compounds and silicon crystals not dissolved under electron-beam processing (Fig. 5.6d).

Fatigue tests conduce to serious transformations of a high-speed crystallization structure. A structure of cellular crystallization remains under a low number of loading cycles (132,000 cycles), but the state of silicon layers is modified significantly. Characteristic microphotographs (Fig. 5.7 and Fig. 5.8) demonstrate that silicon layers are split into separately arranged particles, the sizes of which vary in a range of 15–30 nm (Fig. 5.7a and Fig. 5.8a). Electron-diffraction micro-pattern obtained on silicon layers display a ring structure (Fig. 5.7d), indicating their nano-structural state, i.e. the formation of separately located nano-dimensional silicon

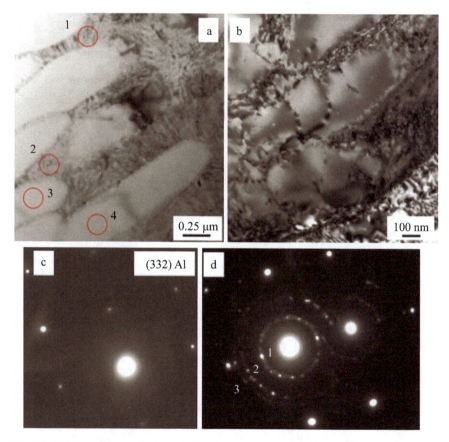

Fig. 5.7 Electron-microscopic view of a silumin structure processed by an intensive electron beam in conditions 20 J/cm², 150 μs, 1 impulse and fractured as a result of 132,000 fatigue loading cycles: (**a**)(**b**) bright fields; (**c**) electron-diffraction micro-pattern obtained on zones 3 and 4 indicated in (**a**); (**d**) electron-diffraction micro-pattern obtained on zones 1 and 2 shown in (**a**); in (**d**) numbers indicate diffraction rings of silicon: 1—(111); 2—(220); 3—(113)

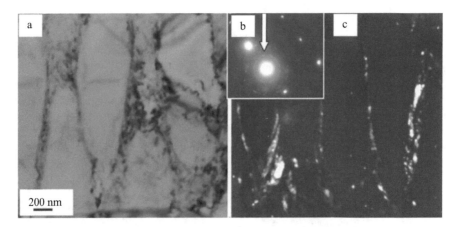

Fig. 5.8 Electron-microscopic view of a silumin structure processed by an intensive electron beam in conditions 20 J/cm^2, 150 µs, 1 impulse and fractured as a result of 132,000 fatigue loading cycles: (**a**) bright field; (**b**) electron-diffraction micro-pattern, arrow indicates a reflex, where a dark-field image was shot, (**c**) dark-field image obtained in a Si reflex [4]

particles. A dislocation substructure represented by separately arranged dislocations is detected throughout cells (Fig. 5.7b). An electron-diffraction micro-pattern obtained on the cells demonstrates their mono-crystalline state (Fig. 5.7c).

A conclusion drawn before that silicon layers do not tend to fragment but are cut into separately arranged particles is verified by diffraction rings of silicon seen in the electron-diffraction micro-pattern (Fig. 5.8b). At this place, an electron-diffraction micro-pattern is a spot diffraction pattern. Nano-dimensional silicon particles are found both on the boundary line of aluminum cells and in their volume (Fig. 5.8c). The latter indicates the removal of atomic-level aluminum or aluminum nano-dimensional particles from layers into the volume of cells.

The development of a dislocation substructure in a form of grids or chaotically arranged dislocations, a scalar density of dislocations $\approx 2 \times 10^{10}$ cm^{-2} (Fig. 5.9), confirms the displacement of dislocations during fatigue testing.

Figure 5.10 and Fig. 5.11 show a structure of the silumin surface after 517,000 loading cycles. As a result of 517,000 loading cycles, a structure develops in the surface layer distantly associated with a structure of high-speed cellular crystallization (Fig. 5.10a). First, layers separating aluminum cells are 2–3 times thicker (Fig. 5.11a). Second, electron-diffraction micro-patterns obtained on the layers are ring diffraction patterns (Fig. 5.10c and Fig. 5.11c). At the same time, electron-diffraction micro-patterns made on a volume of cells are spot diffraction patterns (Fig. 5.11b). In addition, layers tend to fragment, i.e. split into disoriented strips (Fig. 5.11d). In a volume of strips a nano-dimensional structure (in a range of 10 nm) substructure is detected (Fig. 5.10b and Fig. 5.11e).

The fracture of a cellular crystallization structure is related to the globularization of silicon particles distributed equally in a volume of grains (Fig. 5.12a) on dislocations or in nodes of dislocation grids (Fig. 5.12b).

5.2 Evolution of Defect Substructure and Phase State of Silumin Irradiated ...

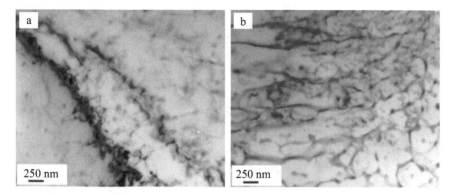

Fig. 5.9 Electron-microscopic view of a silumin dislocation substructure processed by an intensive electron beam in conditions 20 J/cm^2, 150 µs, 1 impulse and fractured as a result of 132,000 fatigue loading cycles

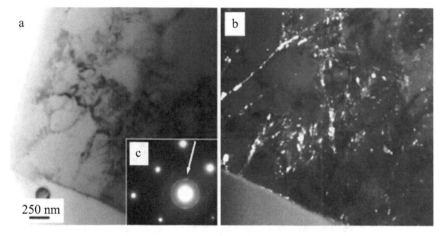

Fig. 5.10 Electron-microscopic view of a silumin structure processed by an intensive pulsed electron beam in conditions 20 J/cm^2, 150 µs, 5 impulses and fractured as a result of 517,000 loading cycles: (**a**) bright field; (**b**) dark field obtained in a reflex of the first diffraction ring (111) Si; (**c**) electron-diffraction micro-pattern, the arrow indicates a reflex where a dark field was obtained (**b**)

High-cycle fatigue tests facilitate the evolution of a dislocation substructure. First, a scalar density of dislocations increases to $\approx 3.4 \times 10^{10}$ cm^{-2}. Second, the rearrangement of dislocations is registered; a structure of chaotically arranged dislocations is replaced by a grid dislocation substructure (Fig. 5.13).

Micro-cracks are found in the subsurface silumin layer exposed to high-cycle (517,000 cycles) fatigue tests. Micro-cracks are detected only in silicon lamellae not dissolved under the irradiation by an electron beam (Fig. 5.14a) or along the boundary between a silicon lamella and an aluminum grain (Fig. 5.14b). These

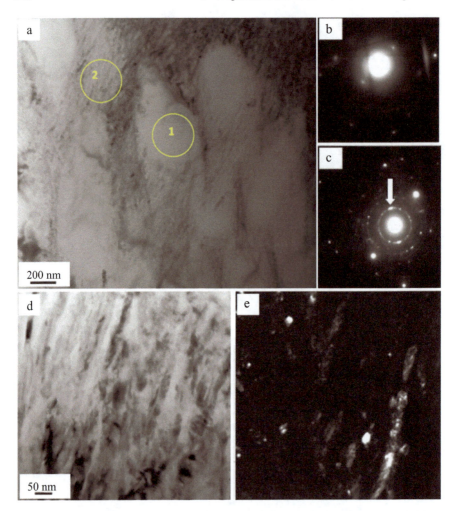

Fig. 5.11 Electron-microscopic view of a silumin structure processed by an intensive electron beam in conditions 20 J/cm^2, 150 µs, 5 impulses and fractured as a result of 517,000 fatigue loading cycles: (**a**)(**d**) bright fields; (**b**)(**c**) electron-diffraction micro-patterns obtained in zones 1 and 2, respectively; (**e**) dark field obtained in a Si reflex [4] (the arrow (**c**) indicates a reflex where a dark field is obtained)

facts also emphasize a negative effect of lamellar silicon inclusions on the fatigue endurance of silumin.

To sum up, a deformation impact typical for the high-cycle fatigue conduces to the nano-structuring of silicon layers on the boundaries of aluminum cells like the observed under 132,000 loading cycles and results in a partial (Fig. 5.10) or full (Fig. 5.12) fracture of a cellular crystallization structure.

The phase composition and the state of the crystal lattice in basic phases of silumin in as-cast state and irradiated by a high-intensity electron beam as well as exposed to

5.2 Evolution of Defect Substructure and Phase State of Silumin Irradiated ... 119

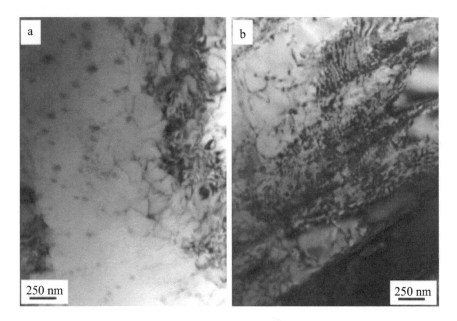

Fig. 5.12 Electron-microscopic view of a silumin structure processed by an intensive electron beam in conditions 20 J/cm^2, 150 μs, 5 impulses and fractured as a result of 517,000 fatigue loading cycles

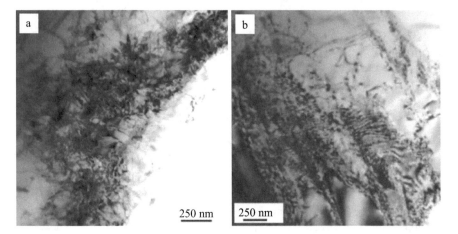

Fig. 5.13 Electron-microscopic view of a silumin dislocation substructure processed by an intensive electron beam in conditions 20 J/cm2, 150 μs, 5 impulses and fractured as a result of 517,000 fatigue loading cycles

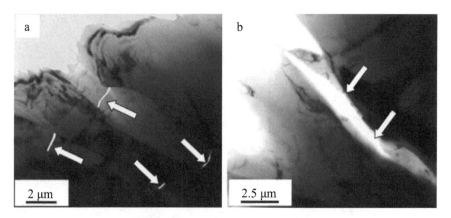

Fig. 5.14 Micro-cracks (indicated by arrows) developing in the surface of silumin processed by an intensive electron beam in conditions 20 J/cm^2, 150 µs, 5 impulses and fractured as a result of 517,000 fatigue loading cycles

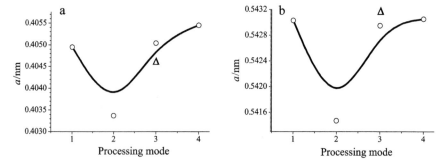

Fig. 5.15 Lattice spacing of aluminum-based (**a**) and silicon-based (**b**) solid solutions vs. processing mode of silumin samples. Impact mode: 1—untreated silumin (as-cast state); 2—as-cast silumin after fatigue tests (130,000 cycles); 3—as-cast irradiated silumin (20 J/cm^2, 150 µs, 1 impulse) and after fatigue tests (132,000 cycles); 4—irradiated as-cast silumin (20 J/cm^2, 150 µs, 5 impulses) and after fatigue tests (517,000 cycles); Δ indicates the lattice spacing in aluminum and silicon for silumin irradiated in conditions 20 J/cm^2, 150 µs, 1 impulse

fatigue tests were examined using the methods of X-ray phase analysis. Figure 5.15 presents a correlation between a lattice spacing of aluminum-based (a) and silicon-based (b) solid solutions and a processing mode (as-cast state and after fatigue failure) of samples.

The results given in Fig. 5.15 allow a conclusion that a lattice spacing of silicon and aluminum exposed to fatigue tests displays maximal changes in untreated silumin (as-cast state). Apparently, the reduction of a lattice spacing in aluminum during fatigue testing of initial samples (Fig. 5.15a, mode 2) is attributed to the dissolution of silicon inclusions and introduction of silicon atoms into the aluminum lattice. A radius of silicon atom is smaller than that of aluminum by 0.0054 nm; therefore, the saturation of an aluminum-based solid solution reduces its lattice spacing.

The irradiation of silumin by an electron beam at an energy density of 20 J/cm^2 causes an insignificant decrease in the aluminum lattice spacing. It is possible due to the dissolution of silicon particles and phases of intermetallic compounds, the saturation of a solid solution by silicon atoms reduces the lattice spacing of aluminum, and manganese atoms with a radius bigger than that of aluminum increase its lattice spacing indeed. Therefore, the electron-beam processing at an energy density of 20 J/cm^2 initiates the dissolution of silicon and magnesium-based phases and the saturation of an aluminum-based solid solution by these elements. The subsequent fatigue testing of silumin causes an additional lattice spacing increase of aluminum, probably due to the leaving of silicon atoms from the aluminum crystal lattice (deformation aging of a material).

To sum up, tribological and strength properties of AK12 silumin exposed to fatigue tests are investigated in this chapter, and the study reveals that an increasing number of fatigue loading cycles has a negative effect on hardness, and increases a wear rate and a friction coefficient.

The research into a structure and phase composition of silumin exposed to the fatigue testing is carried out, suggesting that the irradiation of silumin by a high-intensity electron beam facilitates the development of a cellular (columnar) structure in the surface layer (an average size of cells 450 nm). The cells are shown to be separated by silicon layers with a thickness of 80 nm.

A conclusion is drawn that high-cycle fatigue tests of silumin irradiated by a high-intensity pulsed electron beam lead to the rearrangement of a cellular crystallization structure, i.e. silicon layers rupture and long (around 250 nm thick) two-phase layers with nano-dimensional (up to 10 nm) silicon particles form and envelop aluminum cells. The research points out micron and submicron silicon lamellae, which fail to dissolve in electron-beam processing, represent a reason for fatigue micro-cracks.

An assumption is made that the fracture of a high-speed crystallization structure developed in electron-beam processing appears to be one of the key factors to worsen strength and tribological properties of the silumin surface layer.

References

1. Zhang, L., Wang, C., Han, L., Dong, C.: Influence of laser power on microstructure and properties of laser clad Co-based amorphous composite coatings. Surf.S Interfaces **6**, 18–23 (2017)
2. Greer, J.R., Nix, W.D.: Nanoscale gold pillars strengthened through dislocation starvation. Phys. Rev. B **73**, 245410 (2006)
3. Nix, W.D., Greer, J.R., Feng, G., Lilleodden, E.T.: Deformation at the nanometer and micrometer length scales: Effects of strain gradients and dislocation starvation. Thin Solid Films **515**, 3152–3157 (2007)
4. Fenner, D.B., Hirvonen, J.K., Demaree, J.D.: Selected topics in ion beam surface engineering. Engineering Thin Films and Nanostructures with Ion Beams (2005)

Chapter 6
Modifying of Titanium VT6 Alloy Surface by Electrical Explosion Alloying

6.1 Electrical Explosion Alloying of Titanium VT6 Alloy by Titanium Diboride

A specific feature of electrical explosion alloying is suggested to be the forming of a surface layer with well-developed topography, e.g. beads, drops, craters. Various defects tend to develop on the surface of titanium alloy VT6 in the process of its electrical explosion alloying (Fig. 6.1). The high-speed hardening of the molten surface layer initiates the formation of micro-cracks which might fragment the surface (Fig. 6.1). Dimensionally, structure elements of the alloyed surface vary in a broad range from hundreds of micrometers (Fig. 6.1a) to tens to hundreds of nanometers (Fig. 6.1c).

An analysis of the alloyed surface structure carried out in backscattered electrons has revealed zones with a significantly different contrast (Fig. 6.1a). This fact indicates a difference in the element composition of these zones, i.e. the brighter the contrast of a backscattered electrons image, the higher the atomic number of elements constituting the material under study.

In fact, X-ray microanalysis of the element composition in titanium surface layer has pointed out the non-uniform distribution of its elements. Data presented in the table (Fig. 6.2) highlight the following characteristics of the developed layer. In the first place, zones are saturated by yttrium atoms (Fig. 6.2, zone 1); secondly, it lacks energy spectra associated with boron atoms; another feature is a substantial amount of oxygen and carbon which indicates an insufficient vacuum level in a process chamber; finally, there is no energy spectra of vanadium atoms. Therefore, as a result of electrical explosion alloying of titanium by a titanium diboride powder, zones with a very different concentration of alloying elements develop in the surface layer [1–8].

A study on the structure of cross-sectional etched micro-sections demonstrates the extraordinary roughness of the modified layer (Fig. 6.3). Arrows in Fig. 6.3 indicate the surface of electrical explosion alloying. An analysis of cross-sectional micro-sections has shown a quite broken boundary separating the alloyed layer (a grey

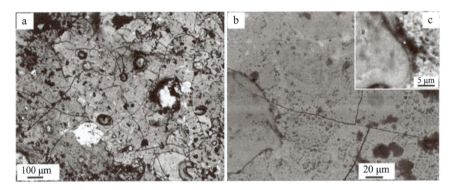

Fig. 6.1 Structure of titanium alloy VT6 surface exposed to electrical explosion alloying by titanium diboride. Backscattered electrons imaging, reprinted from ref. [1] (Copyright 2014, with permission from AIP Conference Proceedings)

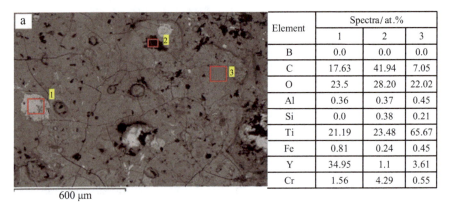

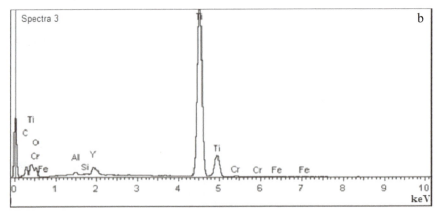

Element	Spectra/at.%		
	1	2	3
B	0.0	0.0	0.0
C	17.63	41.94	7.05
O	23.5	28.20	22.02
Al	0.36	0.37	0.45
Si	0.0	0.38	0.21
Ti	21.19	23.48	65.67
Fe	0.81	0.24	0.45
Y	34.95	1.1	3.61
Cr	1.56	4.29	0.55

Fig. 6.2 Structure of titanium alloy VT6 surface exposed to electrical explosion alloying by titanium diboride. (**a**) structure of titanium alloy VT6 surface; (**b**) energy spectrum obtained in zone 3; the table provides the data of X-ray microanalysis of zones indicated in (**a**)

6.1 Electrical Explosion Alloying of Titanium VT6 …

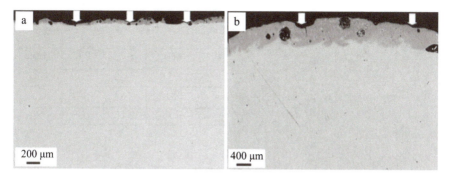

Fig. 6.3 Morphology of titanium alloy VT6 exposed to electrical explosion alloying by titanium diboride

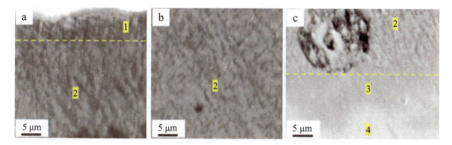

Fig. 6.4 Micro-section structure of a compound based on titanium alloy VT6 exposed to electrical explosion alloying by titanium diboride

surface layer, Fig. 6.3b) and the main material in a sample. Apparently, adhesion (a level of cohesion) of the modified layer and the matrix is high.

Examining the morphology, an alloyed material can be divided at least into four layers: a surface (Fig. 6.4a, layer 1), a border-line layer (Fig. 6.4a, b, layer 2), an intermediate layer (Fig. 6.4c, layer 3), and a heat impact layer (Fig. 6.4c, layer 4), penetrating gradually into a main material of the sample.

Layer 1 and layer 2 principally demonstrate a needle-shaped structure. Dimensionally, a needle-shaped structure in layer 1 is smaller than in layer 2. The structure is unfound in layer 3, evidencing tiny sizes of its elements.

A difference in the etching ability of the layers above suggests their various phase and element compositions. The element composition of the layers was examined in X-ray microanalysis. Figure 6.5 instances an X-ray microanalysis of the alloyed layer.

Several conclusions may be deduced from data provided in the table (Fig. 6.5). In the first place, lacking energy spectra of boron atoms suggest a low concentration of this element in the layer under study. Secondly, a high concentration of carbon atoms evidences a process chamber is pollution by this element. Another fact is the absence

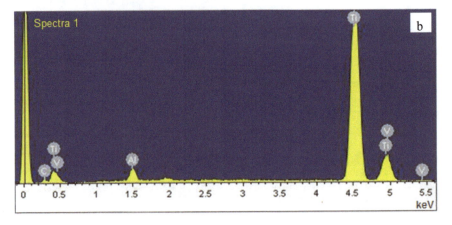

Fig. 6.5 Micro-section structure of an alloy based on titanium alloy VT6 exposed to electrical explosion alloying by titanium diboride. (**a**) structure of titanium alloy VT6 surface; (**b**) energy spectrum obtained in zone 1; the table provides the data of X-ray microanalysis of zones indicated in (**a**)

of yttrium on the alloyed surface (Fig. 6.1). Finally, a concentration of vanadium and aluminum atoms changes regularly with a distance from the alloyed surface.

Interestingly, roundish zones, the structure and element composition of which differ from the surrounding material (Fig. 6.5, zone 4), are detected in the alloyed layer (Fig. 6.3). These zones appear to be pores generating in high-speed melting of the material in a plasma flow. In preparing micro-sections, pores are opened and saturated by elements of a grinding paste.

6.2 Electrical Explosion Alloying of Titanium Alloy VT6 Surface by Boron Carbide

Figure 6.6 provides images of titanium alloy VT6 surface structure developed as a result of electrical explosion alloying by boron carbide. The image demonstrates well-developed topography. Elements of surface structure have various sizes—from hundreds of micrometers (Fig. 6.6a) to tens to hundreds of nano-meters (Fig. 6.6b, c).

The pattern of alloying elements in the processed layer is non-uniform. This assumption is corroborated by a study of the alloyed surface in backscattered electrons. A black and white contrast (Fig. 6.7a) highlights zones saturated with a heavy metal (titanium) and quite light elements (carbon and boron). A qualitative study of the alloying element pattern on the processed surface was carried out by the methods of X-ray microanalysis with the focus on the differentiating element composition in various zones. Data summarized in the table (Fig. 6.7) demonstrate a heterogeneity coefficient of alloying elements in the surface (an aggregate amount of boron, carbon, and oxygen in bright and dark fields) to range up to 2.8. Therefore, zones with a more than 2.5 times difference in the concentration of alloying elements develop as a result of electrical explosion alloying of titanium by a boron carbide powder.

The research into the structure of cross-sectional micro-sections has pointed out that the thickness of the modified layer and the pattern of the alloying elements are

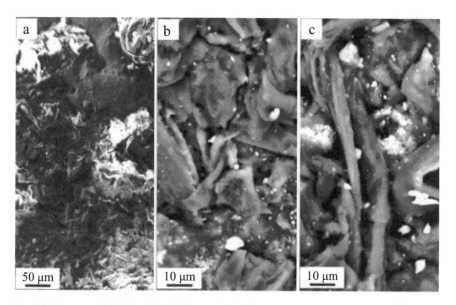

Fig. 6.6 Structure of titanium alloy VT6 exposed to electrical explosion alloying by boron carbide, reprinted from ref. [4] (Copyright 2014, with permission from Begell House Inc., Publishers)

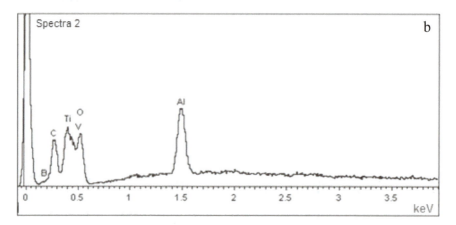

Fig. 6.7 X-ray microanalysis of titanium alloy VT6 exposed to electrical explosion alloying by boron carbide. (**a**) structure of titanium alloy VT6 surface; (**b**) energy spectrum obtained in zone 2; the table provides the data of X-ray microanalysis of zones indicated in (**a**), reprinted from ref. [4] (Copyright 2014, with permission from Begell House Inc., Publishers)

highly heterogeneous. A total thickness of the alloyed layer is measured to be 10–50 μm (Fig. 6.8) owing to the well-developed topography. Arrows (Fig. 6.8) indicate the surface of electrical explosion alloying.

A criterion of morphology (a degree of etching ability) in the alloyed material allows the singling out of minimum 4 layers (numbered in Fig. 6.9b): 1—a surface layer; 2—a border-line layer; 3—an intermediate layer; 4—a heat impact layer, penetrating gradually into a main material of the sample (not shown in the figure). The etching ability was determined to be different not only in layers but also within one layer. The singled out layers possess a certain substructure the elements of which are in a range of 1 μm (Fig. 6.9c, d). A diversity of the etching ability in the layers above emphasizes their different phase and element composition. The element composition of the layers was examined using the methods of X-ray microanalysis. Figure 6.10 instances the element composition of an alloyed layer.

6.2 Electrical Explosion Alloying of Titanium Alloy VT6 ...

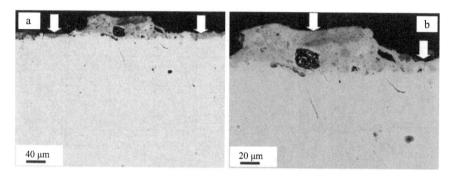

Fig. 6.8 Morphology of titanium alloy VT6 exposed to electrical explosion alloying by boron carbide

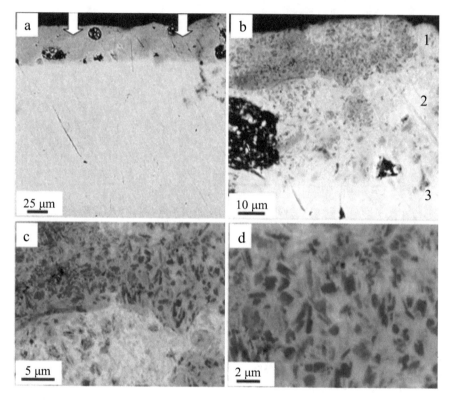

Fig. 6.9 Micro-section structure of a compound based on titanium alloy VT6 exposed to electrical explosion alloying by boron carbide

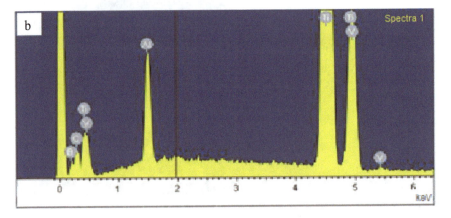

Fig. 6.10 X-ray microanalysis of titanium alloy VT6 micro-section exposed to electrical explosion alloying by boron carbide. (**a**) structure of titanium alloy VT6 surface; (**b**) energy spectrum obtained in zone 1; the table provides the data of X-ray microanalysis of zones indicated in (**a**), reprinted from ref. [4] (Copyright 2014, with permission from Begell House Inc., Publishers)

Assessing the data in the table (Fig. 6.10), an arrangement regularity of alloying elements (boron and carbon) can be found out. The concentration of boron and carbon drops as a distance from the alloyed surface increases (spectra 1–3). Sporadically, roundish zones are detected in the alloyed layer (Fig. 6.10, zone 4). Their structure and element composition (principal detected elements are carbon, oxygen, and silicon) are unlike the surrounding material (Fig. 6.10, table, spectrum 4). These zones are assumed to develop owing to the mechanical grinding of the material under study. It is possible because particles of a grinding paste penetrate into the surface of a micro-section and further the etching of a prepared micro-section.

6.3 Electrical Explosion Alloying of Titanium Alloy VT6 Surface by Silicon Carbide

Figure 6.11 demonstrates a structure of titanium alloy VT6 surface carbonized in electrical explosion with a weighted portion of silicon carbide powder. The processing gives rise to highly rough (Fig. 6.11a) and well-developed topography of the surface. A diversity of structure elements is detected: beads caused by the radial flow of a metal (Fig. 6.11b); micro-cracks and micro-craters (Fig. 6.11c); micro-pores (Fig. 6.11d);

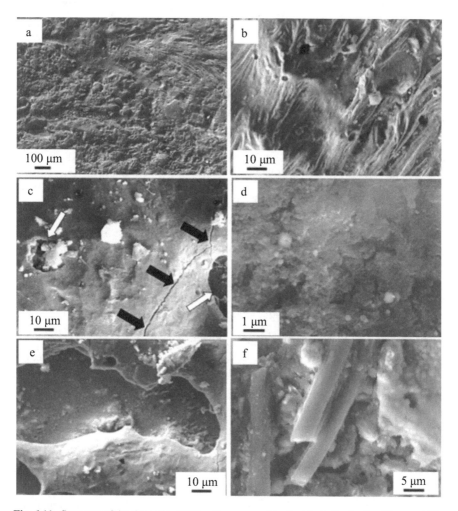

Fig. 6.11 Structure of titanium alloy VT6 surface exposed to electrical alloying by silicon carbide: (**a**) Structure of the surface. Dark arrows indicate micro-cracks; light arrows show micro-craters; (**b**) beads; (**c**) micro-cracks and micro-craters; (**d**) micro-pores; (**e**) overlayers; (**f**) rods

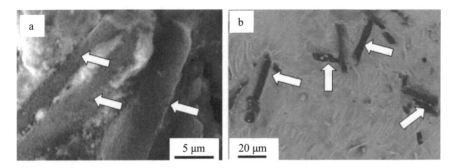

Fig. 6.12 Structure of titanium alloy VT6 surface exposed to electrical alloying by silicon carbide: (**a**) secondary electrons; (**b**) backscattered electrons. Arrows indicate fragments of graphitized carbon fiber

overlayers—particles of the powder or fragments of graphitized carbon fiber penetrating into the melt, facilitating its splashing out (Fig. 6.11e); rods of various sizes (Fig. 6.11f).

Fragments of the exploded graphitized carbon fiber and particles of the weighted powder portion are found in the alloyed layer. They are esteemed to bring about the highly inhomogeneous distribution of carbon and silicon on the processed surface. An SEM-based analysis in backscattered electrons of the surface structure (Figs. 6.11f and 6.12a) shows rods with a length of 10–60 μm and a thickness in a range of 2–5 μm.

The backscattered electrons imaging (Fig. 6.12b) reveals zones of the surface with a considerably different contrast, where a dominant part is grey and rods are dark-colored. When imaged in backscattered electrons, zones of the material saturated by atoms with a higher atomic weight are brighter, whereas zones containing atoms with a relatively low atomic weight are dark-colored. In this study, titanium is an element with a higher atomic weight; therefore, the surface layer is formed by atoms of titanium. Carbon has a relatively low atomic weight, so rods (Fig. 6.12b) represent the fragments of graphitized carbon fiber as indicated by arrows in the figure.

The X-ray microanalysis has pointed out that alloying elements are not distributed uniformly in the surface of titanium alloy VT6 in the process of electrical explosion alloying. The study of energy spectra (Fig. 6.13b, d) highlights silicon-saturated zones in the surface layer of titanium alloy VT6. The whole subsurface layer of the irradiated material is alloyed by silicon in a significantly lower amount (Fig. 6.13c, d).

To sum up, the data of X-ray microanalysis demonstrate that electrical explosion alloying of alloy VT6 modifies the surface and forms a coating with a quite heterogeneous distribution of alloying elements. Zones saturated with silicon atoms (probably conglomerates of silicon carbide powder particles) and zones containing carbon atoms (fragments of graphitized carbon fiber) are detected.

6.3 Electrical Explosion Alloying of Titanium Alloy … 133

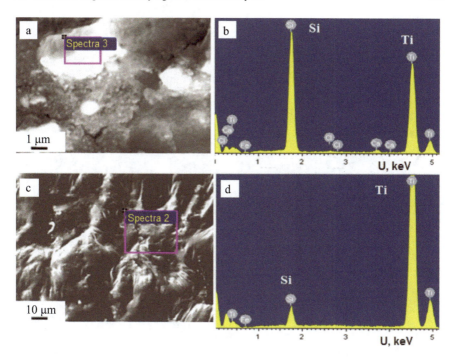

Fig. 6.13 Structure of titanium alloy VT6 surface exposed to electrical alloying by silicon carbide: (**a**) (**c**) structure of the surface; (**b**) (**d**) energy spectra

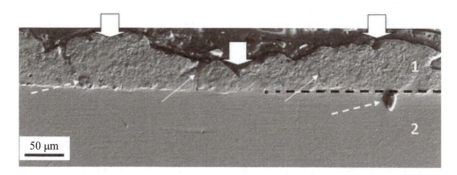

Fig. 6.14 Cross-section of titanium alloy VT6 exposed to electrical explosion alloying by silicon carbide. 1—layer of electrical explosion alloying; 2—main volume of the material

To research a state of the layer after the electrical explosion alloying by boron carbide, cross-sectional micro-sections were used. Figure 6.14 shows a typical structure of a titanium alloy VT6 micro-section. White arrows indicate the surface of alloying. Continuous thin arrows point at micro-cracks and micro-craters, and thin dotted arrows show macro and micro-pores. The surface of the modified material is

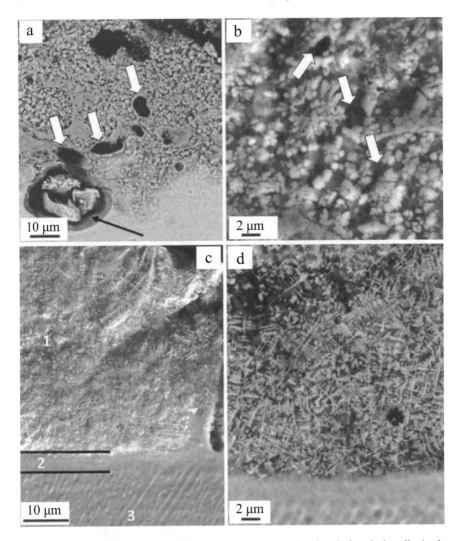

Fig. 6.15 Structure of titanium alloy VT6 cross-section exposed to electrical explosion alloying by silicon carbide: (**a**) (**b**) Structure of the cross-section. White arrows indicate fragments of graphitized carbon fibers. The dark arrow shows macro and micro-pores; (**c**) (**d**) dendrite structure of the cross-section

highly rough as seen in the cross-sectional micro-section of the alloyed layer (layer 1).

Figure 6.15a, b demonstrate a comprehensive analysis of the cross-sectional micro-section, which indicates fragments of graphitized carbon fiber and particles of the weighted powder portion (bright arrows) as well as macro and micro-pores, and cavities (dark arrows) throughout the surface layer.

Another feature of the alloyed layer—a highly heterogeneous structure is established. The study found out zones of the alloyed layer with a dominating globular structure (Fig. 6.15a, b) and zones with a dendrite structure (Fig. 6.15 c, d).

To conclude, electrical explosion alloying of the titanium alloy VT6 surface facilitates the development of a highly heterogeneous structure, element, and phase composition. Summarizing the studies reviewed above, conclusions are drawn [1–8].

1. As a result of electrical explosion alloying of titanium alloy VT6 by titanium diboride powder, a modified layer with a thickness up to 50 μm is formed, and it demonstrates a high level of roughness and contains micro-cracks, pores, etc. Zones containing yttrium atoms are detected. It is reported on a significant difference in alloying elements as well as lacking energy spectra of boron and vanadium atoms which are components of titanium alloy VT6. The high-speed melting of the material in a plasma flow initiates the development of pores on the surface.
2. The modified layer with a thickness varying in a range from 10 to 50 μm is a product of electrical explosion alloying of titanium alloy VT6 by boron carbide powder. Zones with a significantly different (more than 2.5 times) concentration of alloying elements develop in the surface layer. Layers with a certain substructure, the elements of which are in a range of 1 μm, are detected in the alloyed material. The concentration of boron and carbon reduces as the distance from the alloyed surface increases.
3. Electrical explosion alloying of titanium alloy VT6 by boron carbide powder results in the formation of a modified layer with a thickness of 40–60 μm. Zones containing silicon (conglomerates of initial silicon carbide particles) and zones saturated with carbon atoms are detected. Atoms of carbon represent fragments of graphitized carbon fibers with rods (10–60 μm long and 2–5 μm thick).
4. An outcome of electrical explosion alloying of titanium alloy VT6 surface by titanium diboride, boron carbide, and silicon carbide represents a gradient structure developing in the processed zone. The element composition of this structure depends on a distance to the alloyed surface. The pattern of elements in the surface layer is inhomogeneous.
5. Electrical explosion alloying of titanium alloy VT6 surface by silicon carbide conduces to the development of a multiphase polycrystalline structure; its grains are composed of sub-micro and nano-scale lamellae with cross-sections of 10–30 nm. Globular particles of the second phases are detected on boundaries and in joints of grains and sub-grains.

References

1. Konovalov, S. et al.: The role of electro-explosion alloying with titanium diboride and treatment with pulsed electron beam in the surface modification of VT6 alloy. In: AIP Conference Proceedings, vol. 1683, 020093 (2015)

2. Ivanov, Y.F., et al.: Modification of the surface of the VT6 alloy by plasma of electric explosion of a conducting material and by electron beam. Russ. J. Non-Ferrous Met. **55**, 51–56 (2014)
3. Ivanov, Y. F. et al.: Surface hardening alloy VT6 of electric explosion and by electron beam. In: AIP Conference Proceedings, vol. 1623, pp. 217–220 (2014)
4. Ivanov, Y.F., Kobzareva, T.A., Gromov, V.E., Budovskikh, E.A., Bashchenko, L.P.: Electroexplosive doping of titanium alloy by boron carbide and subsequent electron beam processing. High Temp. Mater. Process. (An Int. Q. High-Technology Plasma Process) **18**, 281–290 (2014)
5. Gromov, V.E. et al.: Modification of the titanium alloy surface in electroexplosive alloying with boron carbide and subsequent electron-beam treatment. In: AIP Conference Proceedings, vol. 1683, 020068 (2015)
6. Gromov, V., Kobzareva, T., Ivanov, Y., Budovskikh, E., Baschenko, L.: Surface modification of Ti alloy by electro-explosive alloying and electron-beam treatment. In: AIP Conference Proceedings, vol. 1698, 030006 (2016)
7. Kobzareva, T.Y., Gromov, V.E., Ivanov, Y.F., Budovskkh, E.A., Konovalov, S.V.: Electroexplosive doping of VT6 titanium alloy surface by boron carbide. IOP Conf. Ser. Mater. Sci. Eng. **150**, 012042 (2016)
8. Konovalov, S.V. et al.: Mathematical model of mass transfer at electron beam treatment. In: AIP Conference Proceedings, vol. 1800, 030009 (2017)

Chapter 7
Modifying of Titanium Alloy VT6 Surface by Electrical Explosion Alloying and Electron-Beam Processing

7.1 Research into Titanium Alloy VT6 Processed in Electrical Explosion of Diboride and Irradiated by Electron Beam

A common characteristic of electrical explosion alloying is reported to be the well-developed surface topography of the modified material. A contrast obtained by the methods of SEM has pointed out that alloying elements distribute dissimilarly on the material surface, especially in a state formed as a result of electrical explosion alloying (Fig. 7.1a). That is, a material with a high content of light elements seems to be darker than a material containing atoms of metals—components of the alloy under study—or introduced into it during alloying.

Once a surface alloyed in the electrical explosion of diboride is irradiated by a high-intensity pulsed electron beam, the topography is transformed considerably and alloying elements are rearranged in the surface layer. For instance, data (Fig. 7.1b, c) demonstrate the smoothing of a surface, a reducing number of microcracks; a black and white contrast in the image of the modified surface is substituted mainly by the grey one. The latter indicates that alloying elements distribute more homogeneously in the micro-section plane owing to the supplementary remelting of the surface in the process of electron-beam irradiation.

The high-speed melting and further rapid self-hardening of the surface due to the heat removal into a cold volume of a sample result both in the smoothing of a surface and significantly transforms the material structure.

Figure 7.2 provides a view of a typical titanium surface developed if exposed to the supplementary electron-beam irradiation of the modified layer (18 keV, $E_S = 50$ J/cm^2, $\tau = 100$ µs, $N = 10$ impulses, 0.3 s^{-1}).

Examining the alloyed surface irradiated by an electron beam, a structure consisting mainly of thin lamellae (shells) was revealed (Fig. 7.2). In most cases, lamellae are regular n-sided polygons (tetragons or hexagons). Cross-sectional sizes of polygons vary in a range up to 5 µm; their thickness is determined to be 0.2–0.3 µm.

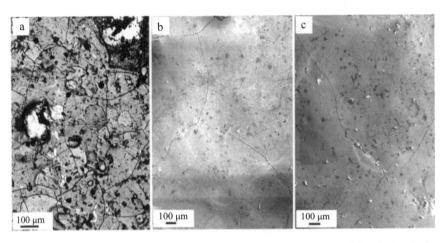

Fig. 7.1 Structure of the processed titanium alloy VT6 surface: (**a**) as a result of electrical explosion alloying by a diboride; (**b**) as a result of electron-beam processing ($E_S = 50$ J/cm^2, $\tau = 100$ μs, $N = 10$ impulses); (**c**) as a result of electron-beam processing ($E_S = 60$ J/cm^2, $\tau = 100$ μs, $N = 10$ impulses)

A porous structure is developed in this case. A structure with lamellae located parallel to the sample surface is less frequent (Fig. 7.2d).

Data (Fig. 7.2e, f) show a regular pattern of lamellae, i.e. lamellae located parallel to the sample surface are growth (crystallization) centers of lamellae perpendicular to the surface. Evidently, crystallization centers will be particles of initial titanium diboride powder.

The irradiation of an alloyed layer by an electron beam (18 keV, $E_S = 50$ J/cm^2, $\tau = 100$ μs, $N = 10$ impulses, 0.3 s^{-1}) fails to homogenize totally the material surface layer. An X-ray microanalysis was carried out locally with the focus on the contrast of surface layer zones. Findings of analysis presented in the table (Fig. 7.3) indicate atoms of yttrium in zones of a clearly brighter contrast (Fig. 7.3, Spectrum 1). The zones with a light-grey contrast contain atoms of boron, carbon, and titanium; therefore, they developed owing to the liquid phase alloying of titanium by elements of plasma and a weighted powder portion (Fig. 7.3, Spectrum 2).

Zones with a darker contrast contain atoms of boron, oxygen, and carbon, being probably agglomerates of initial titanium diboride powder which was not sprayed in the electrical explosion (Fig. 7.3, Spectrum 4).

A raise in energy density of an electron beam up to $E_S = 60$ J/cm^2 hardly causes a significant change of the structure as compared with a structure formed when irradiated by an electron beam with an energy density $E_S = 50$ J/cm^2. It is reported a thin lamellae structure to be generated on a large scale (Fig. 7.4a, b). A new element of the structure is suggested to be the fragmentation of a lamellae structure, i.e. the dividing into zones ≈ 10 μm (Fig. 7.4c).

An X-ray microanalysis of the titanium alloy VT6 surface alloyed in the electrical explosion and irradiated by a high-intensity electron beam ($E_S = 60$ J/cm^2, $\tau = $

7.1 Research into Titanium Alloy VT6 Processed in Electrical Explosion …

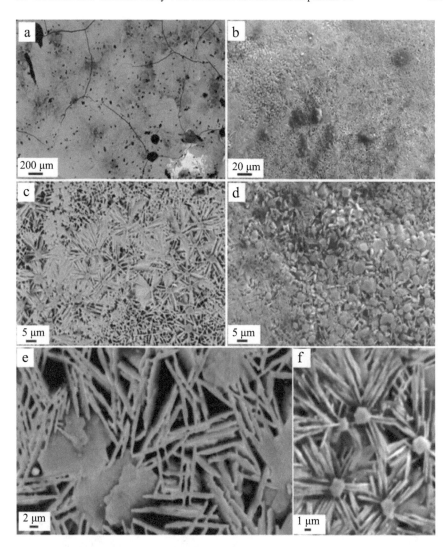

Fig. 7.2 Structure of titanium alloy VT6 exposed to the electrical explosion alloying by diboride and electron-beam processing ($E_S = 50$ J/cm^2, $\tau = 100$ µs, $N = 10$ impulses), reprinted from ref. [1] (Copyright 2014, with permission from AIP Conference Proceedings)

100 µs, $N = 10$ impulses) was conducted according to a contrast level of surface zones (Fig. 7.5a) [2–8].

Grey contrast zones, being dominant elements of the structure, contain atoms of boron, carbon, and titanium (Fig. 7.5, Spectrum 1). In zones of a lighter contrast (Fig. 7.5, Spectrum 4), there are atoms of the initial material and carbon. Dark contrast zones (Fig. 7.5, Spectrum 3) are principally made of atoms of boron, carbon, and oxygen. Bright light contrast zones (Fig. 7.5, Spectrum 2) are enriched by atoms

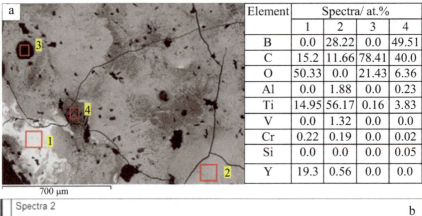

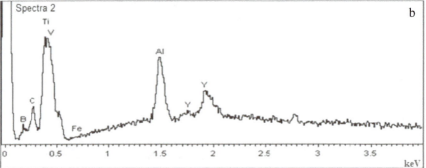

Fig. 7.3 Structure of titanium alloy VT6 surface processed in the electrical explosion of diboride and irradiated by an electron beam ($E_S = 50$ J/cm^2, $\tau = 100$ μs, $N = 10$ impulses). (**a**) structure of the surface; (**b**) energy spectrum obtained on zone 2; the table provides results of an X-ray microanalysis of zones indicated in (**a**)

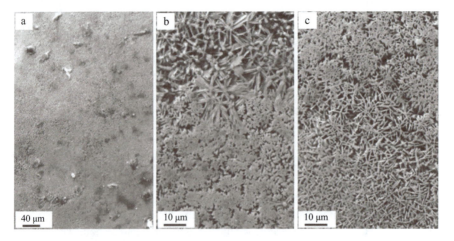

Fig. 7.4 Structure of titanium alloy VT6 surface processed in the electrical explosion of diboride and irradiated by an electron beam ($E_S = 60$ J/cm^2, $\tau = 100$ μs, $N = 10$ impulses)

7.1 Research into Titanium Alloy VT6 Processed in Electrical Explosion ...

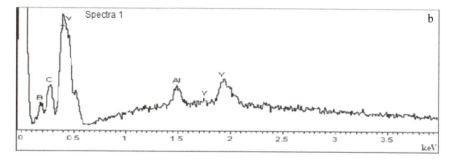

Element	Spectra/ at.%			
	1	2	3	4
B	35.71	13.23	20.33	0.0
C	15.66	17.23	34.11	7.64
O	0.0	34.96	21.66	0.0
Al	0.53	0.81	0.27	6.75
Ti	46.71	21.27	22.89	82.38
V	0.84	0.43	0.42	3.23
Cr	0.0	0.46	0.0	0.0
Si	0.0	0.0	0.33	0.0
Y	0.55	11.59	0.0	0.0

Fig. 7.5 Structure of titanium alloy VT6 surface processed in the electrical explosion of diboride and irradiated by an electron beam ($E_S = 60$ J/cm^2, $\tau = 100$ μs, $N = 10$ impulses). (**a**) structure of the surface; (**b**) energy spectrum obtained on zone 2; the table provides results of an X-ray microanalysis of zones indicated in (**a**)

of yttrium. To conclude, an increase in an electron-beam energy density of $E_S = 60$ J/cm^2 can't result in the full homogenization of the surface layer similarly to the irradiation of the alloyed surface by an electron beam with an energy density of $E_S = 50$ J/cm^2: zones containing atoms of light elements (boron, oxygen, carbon) as well as zones with no alloying elements are still observed.

Transformations in the modified layer structure were examined on etched cross-sectional micro-sections. Figure 7.6 provides characteristic views of cross-sectional micro-sections in samples of titanium alloy VT6 processed in the electrical explosion and irradiated by an electron beam. As seen, the supplementary irradiation by an electron beam improves the roughness and increases the modified layer thickness.

Further, a comprehensive study on the evolution of a cross-section structure in the modified layer is provided. Characteristic views of a cross-section structure in the micro-section of a titanium alloy VT6 sample processed in the electrical explosion and irradiated by an electron beam (an energy density $E_S = 50$ J/cm^2) are given in Fig. 7.7.

It is apparent that the supplementary electron-beam processing at this energy density initiates the formation of a multilayered structure. It is reported on at least

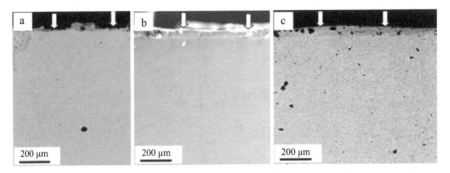

Fig. 7.6 Structure of titanium alloy VT6 surface exposed to a complex treatment. Arrows indicate the processed surface. (**a**) as a result of electrical explosion alloying by diboride; (**b**) as a result of electron-beam processing in conditions $E_S = 50$ J/cm^2, $\tau = 100$ μs, $N = 10$ impulses; (**c**) as a result of electron-beam processing in conditions $E_S = 60$ J/cm^2, $\tau = 100$ μs, $N = 10$ impulses

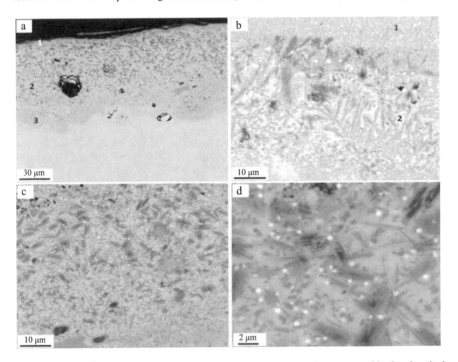

Fig. 7.7 Structure of titanium alloy VT6 cross-sectional micro-section processed in the electrical explosion by titanium diboride and irradiated by an electron beam ($E_S = 50$ J/cm^2, $\tau = 100$ μs, $N = 10$ impulses)

four layers: a surface (1), a border-line layer (2), an intermediate layer (3), and a heat impact layer (4) (Fig. 7.17). The surface and the border-line layer display a needle-shaped (lamellar) structure (Fig. 7.7b–d) and a various dispersion degree. For instance, the structure of the surface is more dispersive (Fig. 7.7b). The structure of the border-line layer is made of lamellar and globular components (Fig. 7.7d).

A similar multilayered needle-shaped (lamellar) structure was also disclosed when the alloyed surface was irradiated by an electron beam ($E_S = 60$ J/cm^2, $\tau = 100$ μs, $N = 10$ impulses [Fig. 7.8]).

Findings of an X-ray microanalysis (Fig. 7.9) demonstrate that the concentration of alloying elements (boron and carbon in this case) regularly decreases with a distance from the surface of alloying. A few atoms of iron and yttrium are detected in the surface layer.

An X-ray microanalysis of the titanium alloy VT6 sample alloyed in the electrical explosion and processed by a high-intensity pulsed electron beam ($E_S = 60$ J/cm^2) has revealed another pattern of alloying elements.

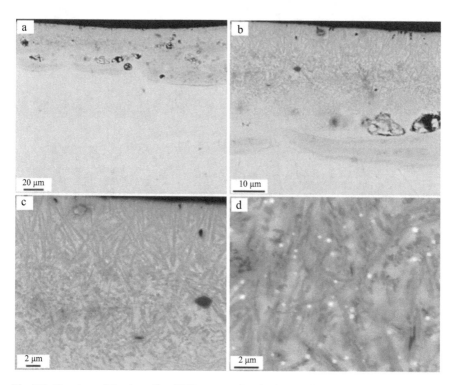

Fig. 7.8 Structure of titanium alloy VT6 cross-sectional micro-section processed in the electrical explosion by titanium diboride and irradiated by an electron beam ($E_S = 60$ J/cm^2, $\tau = 100$ μs, $N = 10$ impulses)

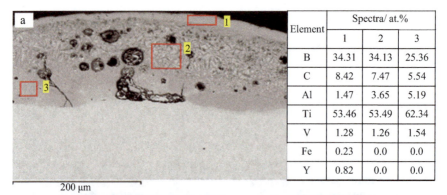

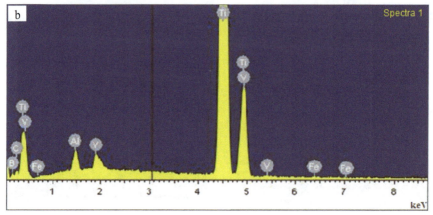

Fig. 7.9 Structure of titanium alloy VT6 surface processed in the electrical explosion of diboride and irradiated by an electron beam ($E_S = 50$ J/cm^2, $\tau = 100$ μs, $N = 10$ impulses). (**a**) structure of the surface; (**b**) energy spectrum obtained on zone 1; the table provides results of an X-ray microanalysis of zones indicated in (**a**)

Data (Fig. 7.10) illustrate that the maximal concentration of boron atoms is in the border-line layer; atoms of carbon are detected throughout the alloyed layer; finally, atoms of yttrium are found both in the surface and in the border-line layer.

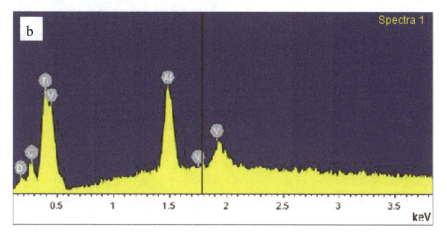

Fig. 7.10 Structure of titanium alloy VT6 surface processed in the electrical explosion of diboride and irradiated by an electron beam ($E_S = 60$ J/cm^2, $\tau = 100$ µs, $N = 10$ impulses). (**a**) structure of the surface; (**b**) energy spectrum obtained on zone 1; the table provides results of an X-ray microanalysis of zones indicated in (**a**)

7.2 Effect of Electron-Beam Processing on Modifying of Titanium Surface Alloyed in an Electrical Explosion by Boron Carbide

As stated above, a distinctive feature of electrical explosion alloying is the well-developed surface topography of a material. Figure 7.11a demonstrates a surface structure of titanium alloy VT6 developed during electrical explosion alloying. This view also indicates the well-developed topography of the processed surface.

A contrast obtained in SEM points at an inhomogeneous pattern of alloying elements in the sample surface. For instance, a material volume with a high content of light elements (carbon and boron) is darker than a volume containing atoms of metals—components of the alloy under study.

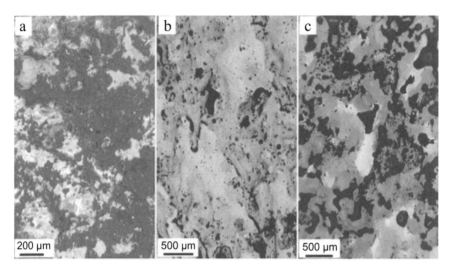

Fig. 7.11 Structure of titanium alloy VT6 surface exposed to a complex treatment: (**a**) as a result of electrical explosion alloying by boron carbide; (**b**) as a result of electron-beam processing ($E_S = 50$ J/cm^2, $\tau = 100$ μs, $N = 10$ impulses); (**c**) as a result of electron-beam processing ($E_S = 60$ J/cm^2, $\tau = 100$ μs, $N = 10$ impulses)

The further irradiation of the modified surface by a high-intensity electron beam changes its topography and distribution of alloying elements in it. That is, the surface is more smoothed; a black and white contrast in the image of the modified surface is substituted mainly by the grey one (Fig. 7.11b, c). The latter indicates a better distribution of alloying elements in the micro-section owing to the electron-beam irradiation.

The high-speed melting and the subsequent fast self-hardening of the surface layer caused by the heat removal into a cold part of the sample smooth the surface and transform a structure of the material. Figure 7.12 provides a characteristic view of a layer structure developed during supplementary electron-beam processing of modified titanium (18 keV, $E_S = 60$ J/cm^2, $\tau = 100$ μs, $N = 10$ impulses, 0.3 s^{-1}). Examining the alloyed surface, two elements of the structure formed as a result of the second treatment were found, i.e. zones with a needle-shaped structure (Fig. 7.13a). The needles range up to 10 μm lengthwise and to 1 μm crosswise. The needles are located mainly perpendicular to the irradiated surface, i.e. in the direction of heat removal. Another typical element of the irradiated surface appears to be relatively smooth zones with elements ranging up to 100 nm (Fig. 7.13b).

Zones presented in Figs. 7.12 and 7.13 differ in the element composition. Figure 7.14 provides data of an X-ray microanalysis, demonstrating core components in zones with a strong dark contrast are alloying elements and oxygen (Fig. 7.14, Spectrum 3). Therefore, they might be formed by particles of the initial powder which fail to dissolve during electrical explosion alloying and further electron-beam processing.

7.2 Effect of Electron-Beam Processing on Modifying of Titanium Surface ...

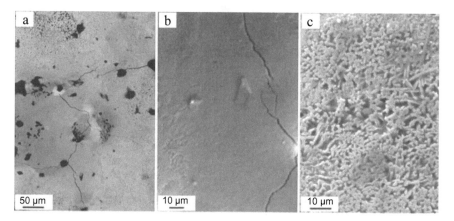

Fig. 7.12 Structure of titanium alloy VT6 surface as a result of electrical explosion alloying by boron carbide and electron-beam processing ($E_S = 50$ J/cm^2, $\tau = 100$ μs, $N = 10$ impulses)

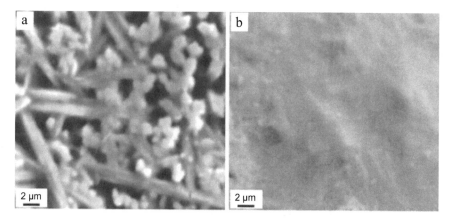

Fig. 7.13 Structure of titanium alloy VT6 surface as a result of electrical explosion alloying by boron carbide and electron-beam processing ($E_S = 50$ J/cm^2, $\tau = 100$ μs, $N = 10$ impulses)

Zones with a nano-dimensional sub-structure are principally formed by atoms of the initial material and a small quantity of carbon atoms (Fig. 7.14, Spectrum 1). As assumed, these zones contain particles of a carbide phase. In needle-shaped zones (Fig. 7.14, Spectrum 2), there are particles of the alloying powder and titanium alloy VT6; so they were formed in the liquid phase alloying of titanium by boron, carbon, and oxygen, i.e. their phase composition might be complex.

The increase in the energy density of an electron beam ($E_S = 60$ J/cm^2) initiates the formation of a prevalently needle-shaped structure (Fig. 7.15). The X-ray microanalysis of needle-shaped zones has pointed out both alloying elements and

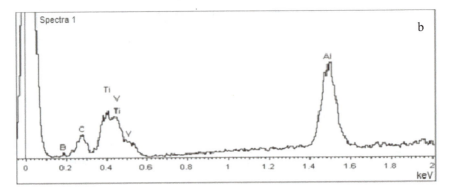

Fig. 7.14 Structure of titanium alloy VT6 surface processed in the electrical explosion of boron carbide and irradiated by an electron beam ($E_S = 50$ J/cm^2, $\tau = 100$ μs, $N = 10$ impulses). (**a**) structure of the surface; (**b**) energy spectrum obtained on zone 1; the table provides results of an X-ray microanalysis of zones indicated in (a), reprinted from ref. [9] (Copyright 2014, with permission from Begell House Inc., Publishers)

components of the initial alloy (Fig. 7.16, Spectrum 1). This evidences that a dissolubility of boron carbide powder in titanium increases as an electron-beam energy density is raised, i.e. a modified surface layer becomes more homogenized.

Transformations in the surface layer structure were investigated using cross-sectional micro-sections. Figure 7.17 shows a characteristic structure of the cross-sectional micro-section in the alloyed layer processed by an electron beam (an energy density $E_S = 50$ J/cm^2).

As seen, the supplementary processing by an electron beam at the given energy density fails to produce a homogenous structure. The surface layer with a thickness of up to 30 μm displays a needle-shaped structure (Fig. 7.17b, c); the sub-structure of a beneath layer is similar to that forming in a material during electrical explosion alloying. Therefore, the electron-beam processing in conditions given above furthers the modification of a layer thinner than 30 μm.

In most cases, a structure of the modified layer is lamellar. Both the study of micro-sections in backscattered electrons (Fig. 7.17a) and the X-ray microanalysis of the

7.2 Effect of Electron-Beam Processing on Modifying of Titanium Surface ...

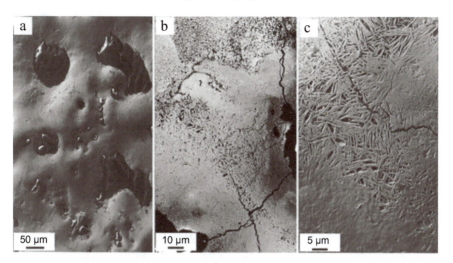

Fig. 7.15 Structure of titanium alloy VT6 surface processed in the electrical explosion of boron carbide and irradiated by an electron beam ($E_S = 60$ J/cm^2, $\tau = 100$ μs, $N = 10$ impulses)

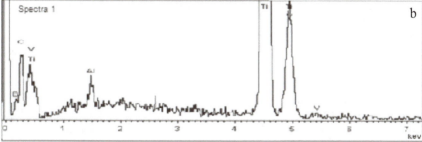

Element	Spectra/ at.%	
	1	2
B	16.54	10.0
C	17.86	72.37
O	0	17.63
Al	0.92	0.0
Ti	62.58	0.0
V	2.11	0.0

Fig. 7.16 Structure of titanium alloy VT6 surface processed in the electrical explosion of boron carbide and irradiated by an electron beam ($E_S = 60$ J/cm^2, $\tau = 100$ μs, $N = 10$ impulses). (**a**) structure of the surface; (**b**) energy spectrum obtained on zone 1; the table provides results of an X-ray microanalysis of zones indicated in (**a**)

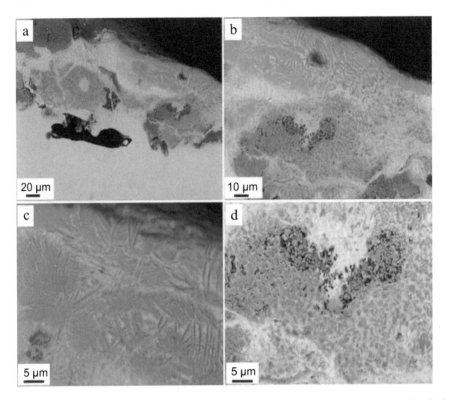

Fig. 7.17 Structure of titanium alloy VT6 cross-sectional micro-section processed in the electrical explosion by boron carbide and irradiated by an electron beam ($E_S = 50$ J/cm^2, $\tau = 100$ μs, $N = 10$ impulses)

element composition emphasize this feature. Layers with a higher and lower content of alloying elements are found. Importantly, a concentration of alloying elements hardly depends on a distance from the alloyed layer to the irradiation surface.

This fact illustrates the development of a multilayered structure in the material: layers with a higher level of alloying (hardened layers) alternate with those of lower alloying (weaker layers).

A similar multilayered structure develops in the modified layer when the material is irradiated by a high-intensity electron beam ($E_S = 60$ J/cm^2, $\tau = 100$ μs, $N = 10$ impulses [Fig. 7.18]). Typical views of cross-sectional micro-sections (Fig. 7.18) demonstrate layers with a different contrast and sub-structure to be thinner than layers which form in the modified layer in the process of electron-beam irradiation ($E_S = 50$ J/cm^2 [Figs. 7.17 and 7.19]).

The pattern of elements determined in an X-ray microanalysis also evidences a layered structure of the surface generated as a result of a treatment combining electrical explosion alloying by boron carbide and high-intensity electron-beam irradiation.

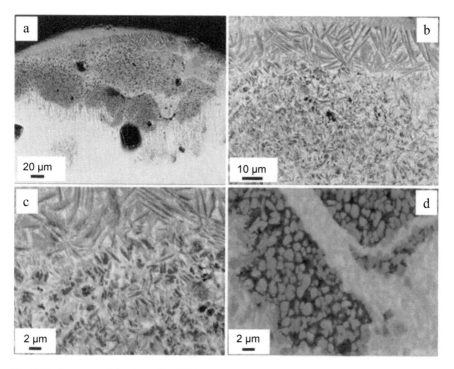

Fig. 7.18 Structure of titanium alloy VT6 cross-sectional micro-section processed in the electrical explosion by boron carbide and irradiated by an electron beam ($E_S = 60$ J/cm^2, $\tau = 100$ µs, $N = 10$ impulses), reprinted from ref. [10] (Copyright 2015, with permission from AIP Publishing)

Data in the table (Fig. 7.20) illustrate a significant difference in concentration of alloying elements in layers with various contrasts. Therefore, the irradiation of the alloyed layer by a high-intensity electron beam results in no homogenization, and instead of it a layered structure is formed. It is assumed that strength and tribological properties of the detected layers will be different due to a diverse concentration of alloying elements in them.

7.3 Effect of Electron-Beam Processing on Modifying of Titanium Surface Alloyed in Electrical Explosion by Silicon Carbide

The electron-beam processing of the surface alloyed in the electrical explosion by a silicon carbide was carried out in conditions: (i) $E_S = 50$ J/cm^2, $\tau = 100$ µs, $N = 10$ impulses; (ii) $E_S = 60$ J/cm^2, $\tau = 200$ µs, $N = 10$ impulses; and the impulse repetition rate is 0.3 s^{-1}.

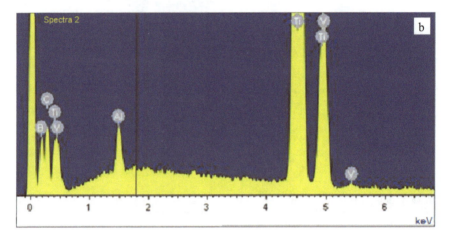

Fig. 7.19 Structure of titanium alloy VT6 surface processed in the electrical explosion of boron carbide and irradiated by an electron beam ($E_S = 50$ J/cm^2, $\tau = 100$ μs, $N = 10$ impulses). (**a**) structure of the surface; (**b**) energy spectrum obtained on zone 2; the table provides results of an X-ray microanalysis of zones indicated in (**a**)

To analyze the irradiated surface, SEM methods were employed. A study has pointed out that the electron-beam homogenizing of samples (for parameters given above) results in serious transformations in surface and subsurface of a sample. As reported, micro-drops, beads disappear in the center of an electron beam affected zone (its diameter expands starting from 10 mm at an energy density of an electron beam $E_S = 50$ J/cm^2 to 18 mm at $E_S = 60$ J/cm^2) a number of micro-cracks decreases significantly and the surface becomes smoother (Fig. 7.21a).

The electron-beam processing of an alloy VT6 surface gives rise to many micro-craters (Fig. 7.21b). A raise in an electron-beam energy density reduces a number of micro-craters enormously, in addition, a study on samples irradiated by an electron beam ($E_S = 60$ J/cm^2) has revealed almost no micro-craters on the processed surface.

Several conclusions might be drawn if a structure forming in the alloy VT6 surface irradiated by an electron beam at an energy density of $E_S = 50$ J/cm^2 is compared with an outcome of irradiation at $E_S = 60$ J/cm^2. A polycrystalline structure with grains in a range from 0.4 to 10 μm (Fig. 7.22a, b) is seen on the processed surface if irradiated

7.3 Effect of Electron-Beam Processing on Modifying of Titanium Surface ...

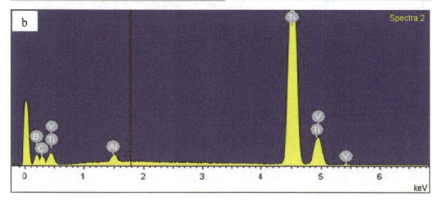

Fig. 7.20 Structure of titanium alloy VT6 surface processed in the electrical explosion of boron carbide and irradiated by an electron beam ($E_S = 60$ J/cm^2, $\tau = 100$ μs, $N = 10$ impulses). (**a**) structure of the surface; (**b**) energy spectrum obtained on zone 2; the table provides results of an X-ray microanalysis of zones indicated in (a), reprinted from ref. [9] (Copyright 2014, with permission from Begell House Inc., Publishers)

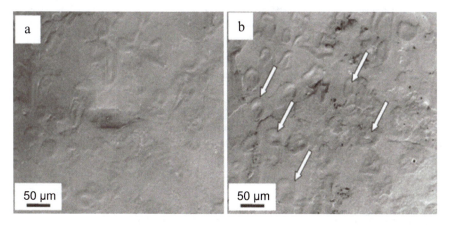

Fig. 7.21 Electron-microscopic view of a titanium alloy VT6 structure processed in the electrical explosion of silicon carbide and irradiated by an electron-beam: (**a**) electron-beam processing $E_S = 50$ J/cm^2, $\tau = 100$ μs, $N = 10$ impulses; (**b**) electron-beam processing in conditions $E_S = 60$ J/cm^2, $\tau = 100$ μs, $N = 10$ impulses. Arrows (**b**) indicate micro-craters

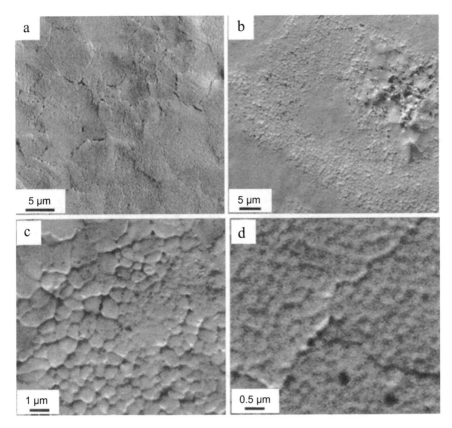

Fig. 7.22 Structure of titanium alloy VT6 surface processed in the electrical explosion by silicon carbide and irradiated by an electron beam ($E_S = 50$ J/cm^2, $\tau = 100$ μs, $N = 10$ impulses). (**a**) polycrystalline structure; (**b**) islands of polycrystalline structure grains; (**c**) dimensions of polycrystalline structure grains; (**d**) high-speed crystallization cells

in conditions $E_S = 50$ J/cm^2, $\tau = 100$ μs, $N = 10$ impulses. Basically, the surface layer is made of grains with an average size of 3 μm. Grains in the untreated material are 8 μm on average. Submicron grains are located like small islands (Fig. 7.22b), their dimensions range from 0.4 to 0.9 μm (Fig. 7.22c). A feature of coarse grains is suggested to be cells of high-speed crystallization found in their volume (Fig. 7.22e). Dimensions of cells vary from 250 to 300 nm.

The electron-beam irradiation of the alloyed surface ($E_S = 60$ J/cm^2, $\tau = 100$ μs, $N = 10$ impulses) results in a two-level grain structure (Fig. 7.23a). Grains of the first (large-scale) level are in a range from 6 to 90 μm, and the average size of the grain is determined to be 28 μm (Fig. 7.23b). Grains of the second scale level are in regions of diverse forms and dimensions, forming long layers (Fig. 7.23c). Their dimensions vary from 0.5 to 2.1 μm, with an average size of 0.9 μm (Fig. 7.23d).

High-speed crystallization cells (300–400 nm) are detected in first-level grains like a structure developing when alloy VT6 is processed by an electron beam in

7.3 Effect of Electron-Beam Processing on Modifying of Titanium Surface …

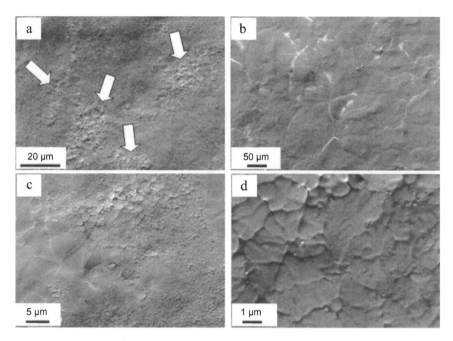

Fig. 7.23 Electron-microscopic view of titanium alloy VT6 structure processed in the electrical explosion of silicon carbide and irradiated by an electron beam ($E_S = 60$ J/cm^2, $\tau = 100$ μs, $N = 10$ impulses). (**a**) two-level grain structure. Arrows indicate zones second-level grains are localized. (**b**) first-level (large-scale). (**c**) second-level grains. (**d**) dimensions of second-level grains

conditions $E_S = 60$ J/cm^2, $\tau = 100$ μs, $N = 10$ impulses (Fig. 7.22d). A distinctive feature of the alloy VT6 structure processed by an electron beam (60 J/cm^2, 100 μs, 10 impulses, 0.3 s^{-1}) is suggested to be zones with a dendrite crystallization structure (Fig. 7.24). This fact illustrates a decrease in the crystallization speed of titanium alloy VT6 once an energy density of an electron beam is raised.

The electron-beam processing of the surface alloyed in the electrical explosion of silicon carbide results in the homogenizing of the surface layer in titanium alloy VT6 and reduces its roughness. The X-ray microanalysis of the element composition reveals no zones containing silicon and carbon.

It is important that the further electron-beam processing of titanium alloy VT6 surface alloyed in the electrical explosion by silicon carbide considerably smoothes the modified surface and optimizes a thickness of the modified layer (Fig. 7.25). A modified layer processed by an electron beam is as thick as 20–30 μm and tends to thin as an energy density of an electron beam goes up. The melting of the modified layer by an electron beam fails to improve the element composition of the alloyed layer because fragments of graphitized carbon fibers and particles of a weighted powder portion penetrate into the melt. Dark arrows (Fig. 7.25) indicate fragments of graphitized carbon particles. In addition, a few micro-pores and micro-cracks are detected in the alloyed layer (indicated by a light arrow in Fig. 7.25).

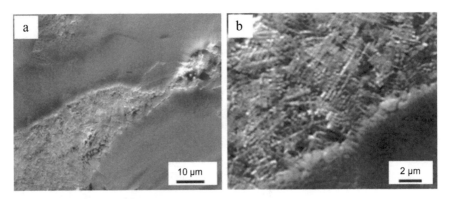

Fig. 7.24 Electron-microscopic view of titanium alloy VT6 structure processed in the electrical explosion of silicon carbide and irradiated by an electron beam ($E_S = 60$ J/cm^2, $\tau = 100$ μs, $N = 10$ impulses): (**a**) dendrite crystallization of the surface layer; (**b**) fist-level (large-scale) grains

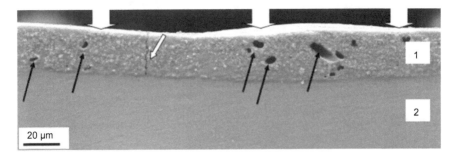

Fig. 7.25 Electron-microscopic view of titanium alloy VT6 cross-section processed in the electrical explosion of silicon carbide and irradiated by an electron beam ($E_S = 60$ J/cm^2, $\tau = 100$ μs, $N = 10$ impulses). Light arrows show the processed surface; thin dark arrows—graphitized carbon fiber particles; thin light arrow—micro-crack; 1—alloyed layer; 2—heat impact layer

The electron-beam processing of the titanium alloy VT6 surface by silicon carbide initiates the melting of the alloyed layer. Any mode of electron-beam processing facilitates the development of a multilayered mainly globular structure. The structure of the surface layer is coarser than in the border-line layer (Fig. 7.26).

To sum up, the combination of the electrical explosion alloying by titanium diboride and electron-beam processing of the surface layer in titanium alloy VT6 furthers the smoothing of the alloyed surface; as a result, a structure made of thin lamellae (cross-sectional dimensions up to 5 μm, a thickness range from 0.2 to 0.3 μm) develops. Lamellae are perpendicular to the surface of a sample and form a porous structure at an electron energy density $E_S = 50$ J/cm^2. Particles of titanium diboride are suggested to be centers of crystallization in this case. A raise in a beam energy density to $E_S = 60$ J/cm^2 brings about the fragmentation of a thin lamellar structure with zones around 10 μm.

7.3 Effect of Electron-Beam Processing on Modifying of Titanium Surface ...

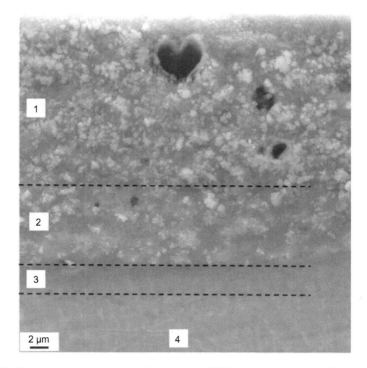

Fig. 7.26 Electron-microscopic view of titanium alloy VT6 cross-section processed in the electrical explosion of silicon carbide and irradiated by an electron beam ($E_S = 60$ J/cm^2, $\tau = 100$ μs, $N = 10$ impulses). 1—surface layer; 2—border-line layer; 3—intermediate layer; 4—heat impact layer

The further electron-beam processing fails to homogenize the surface entirely: zones containing light elements (boron, oxygen, carbon) and zones with no alloying elements are detected. The regularities of the multilayered structure development have been revealed. The structure of the surface layer is more dispersive at $E_S = 50$ J/cm^2, and the border-line layer is made of lamellar and globular elements. A lamellar structure develops mainly at $E_S = 60$ J/cm^2.

At $E_S = 50$ J/cm^2, the concentration of alloying elements (boron and carbon) drops with a distance from the alloyed surface, and atoms of iron and yttrium are detected in the surface layer. A maximal concentration of boron is found in the border-line layer at $E_S = 60$ J/cm^2; atoms of carbon are detected throughout the alloyed layer; and yttrium in the surface and the border-line layer.

The electron-beam processing of the titanium alloy VT6 surface layer alloyed in the electrical explosion of boron carbide results in a nano-dimensional sub-structure formed by atoms of the initial material and carbon. Electron-beam processing carried out at $E_S = 50$ J/cm^2 leads to the development of needle-shaped zones on the surface of titanium alloy VT6, and the needles are in a range up to 10 μm lengthwise and around 1 μm crosswise. The needles are detected mainly perpendicular to the surface, i.e. in the direction similar to the heat removal. Smooth zones formed by boron carbide powder are also revealed, and elements in these zones range up to 100 nm. A raise in an electron-beam energy density to $E_S = 60$ J/cm^2 furthers the development of a needle-shaped structure and increases the dissolution level of boron carbide particles.

The electron-beam processing of the titanium alloy VT6 surface layer alloyed in the electrical explosion of silicon carbide smoothes the surface topography, reducing the number of micro-cracks, micro-drops, and beads.

The layer modified in the electron-beam processing by silicon carbide is measured to be 20–30 μm and thins if an electron-beam energy density is raised. Any mode of the electron-beam processing causes the development of a multilayered globular structure. The structure of the surface layer is coarser than in the border-line layer. As a result of the electron-beam processing at $E_S = 50$ J/cm^2, a polycrystalline structure forms in the surface (dimensions of grains: 0.4–10 μm). The surface layer is principally made of grains of 3 μm on average, whereas an average size of grains in the material volume is 8 μm. Submicron grains (0.4–0.9 μm) are localized.

A raise in a beam energy density to $E_S = 60$ J/cm^2 results in a two-level grain structure. First-level grains are 6–90 μm, being 28 μm on average. Second-level grains (0.5–2.1 μm) form long layers, being 0.9 μm on average. Zones with a structure of dendrite crystallization are suggested to represent a feature of the structure.

The study has pointed out that the electron-beam processing of the titanium alloy VT6 surface alloyed in the electrical explosion of titanium diboride smoothes the alloyed surface and generates a multilayered structure. The concentration of alloying elements varies regularly in this structure. High speeds of cooling initiated by the pulsed electron-beam processing facilitate the development of a sub-micro and nano-dimensional, as a consequence, the enhancement of strength and tribological properties of the modified material might be expected.

References

1. Konovalov, S. et al.: The role of electro-explosion alloying with titanium diboride and treatment with pulsed electron beam in the surface modification of VT6 alloy. In: AIP Conference Proceedings, vol. 1683, 020093 (2015)
2. Cui, W.: A state-of-the-art review on fatigue life prediction methods for metal structures. J. Mar. Sci. Technol. **7**, 43–56 (2002)
3. Stephens, R.I., Fatemi, A., Stephens, R.R., Fuchs, H.O.: Metal Fatigue in Engineering. Wiley Interscience, Hoboken, NJ (2000)
4. Schutz, W.: A history of fatigue. Eng. Fract. Mech. **54**, 263–300 (1996)
5. Bhaduri, A.: Fatigue. In: Mechanical Properties and Working of Metals and Alloys. Springer Series in Materials Science, vol. 264. Springer, Singapore (2018)
6. Yokobori, T.: Physics of Strength and Plasticity. MIT Press, Boston (1969)
7. Kennedy, A.J.: Processes of Creep and Fatigue in Metals. Oliver & Boyd Ltd., London (1962)
8. Golovin, S.A., Puskar, A.: Microplasticity of a low-carbon steel at low and high loading frequencies. Kov. Mater. **16**, 426–437 (1978)
9. Ivanov, Y.F., Kobzareva, T.A., Gromov, V.E., Budovskikh, E.A., Bashchenko, L.P.: Electroexplosive doping of titanium alloy by boron carbide and subsequent electron beam processing. High Temp. Mater. Process. (An Int. Q. High-Technology Plasma Process) **18**, 281–290 (2014)

10. Gromov, V.E. et al.: Modification of the titanium alloy surface in electroexplosive alloying with boron carbide and subsequent electron-beam treatment. In: AIP Conference Proceedings, vol. 1683, 020068 (2015)

Chapter 8
Microhardness and Wear Resistance of Modified Layers

8.1 Depthwise Distribution of Microhardness in Modified Layers

8.1.1 Role of Powder Portion Weight for Depthwise Microhardness Distribution in the Zone of Electrical Explosion Alloying

Electrical explosion alloying by a weighted TiB_2 powder portion depthwise distribution of microhardness in the alloyed zone, as shown in Fig. 8.1, tends to form two layers. Remaining relatively at the same level, it varies non-monotonously in the first (surface) layer. Once a powder portion weight is doubled (from 60 to 120 mg), the microhardness will increase by 1.3 times (from 900 to 1200 HV), being, therefore, 2.5–3.5 higher than that of as-cast alloy VT6 (350 HV). At the same time, a surface layer with a high microhardness is nearly twice as thick—in a range from 15–17 to 30–35 μm. The microhardness of the second layer (intermediate layer) decreases monotonously. The thickness of the intermediate layer is similar in both cases (15–17 μm).

This tendency of microhardness is related to hardening particles of weighted powder portion—their concentration is considerably higher in the surface layer than in the intermediate one due to the structure of multiphase plasma jet affecting the surface to be hardened. As known, a plasma jet comprises a vapor and plasma front and a relatively slow rear part which contains condensed particles of an exploded conductor and powder portions introduced into the explosion zone.

Electrical explosion alloying by a weighted B_4C powder portion. Similarly to the experiments described above, two layers are found if a weighted portion of boron carbide powder is used in electrical explosion alloying (Fig. 8.2). The thickness of the surface layer is determined to be 20–25 μm and that of the intermediate layer is 20 μm. An eightfold raise of a powder B_4C portion weight (from 60 to 500 mg)

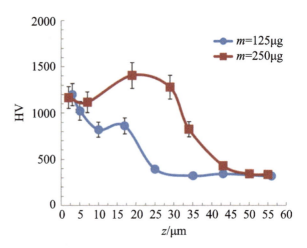

Fig. 8.1 Depthwise distribution of microhardness in a zone of electrical explosion alloying by a TiB$_2$ powder portion with a varying weight

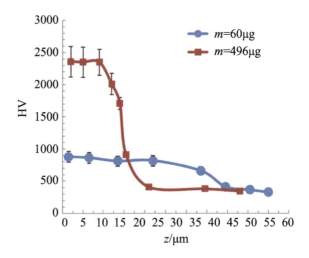

Fig. 8.2 Depthwise distribution of microhardness in a zone of electrical explosion alloying by a B$_4$C powder portion with a varying weight

brings about a 2.8 times increase in the surface layer microhardness (from 850 to 2400 HV).

Electrical explosion alloying by a weighted SiC powder portion. During electrical explosion alloying by a weighted portion of SiC powder, the maximum of microhardness (1100–1200 HV) is recorded in the subsurface layer (at a depth of about 10 μm beneath the surface), and it exceeds the microhardness of the main material volume by 5.5–6 times. The thickness of the subsurface layer is in the range of 10–15 μm. From that time the microhardness drops to 700–800 HV (Fig. 8.3). The total thickness of the hardened layer is estimated to be 25–30 μm. At this place, a second maximum of microhardness is detected at a depth of about 30 μm, i.e. at a depth of the intermediate layer. The microhardness profile varies non-monotonously: the

Fig. 8.3 Depthwise distribution of microhardness in a zone of electrical explosion alloying by a SiC powder portion (50 mg)

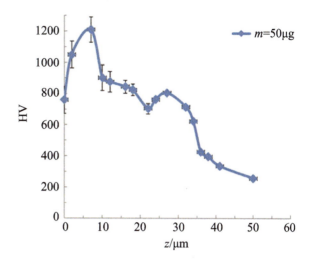

subsurface maximum of microhardness is followed by a lengthy layer with a thickness of 20–25 μm and high microhardness (≈ 800 MPa). The microhardness of the modified layer tends to reduce significantly with a distance from the surface of electrical explosion alloying [1–7].

8.1.2 Importance of Surface Energy Density for Depthwise Microhardness Distribution in Zone Irradiated by Electron Beams

Electron-beam processing of the alloyed surface by a weighted portion of TiB_2 powder. The surface zone alloyed in the electrical explosion of a weighted portion of TiB_2 powder was irradiated by an electron beam in two modes with different surface energy densities. In both cases, the microhardness of the surface layer increases from 1200 to 1800 HV; that is almost five times higher than in as-cast alloy (Fig. 8.4). Interestingly, its depth is similar to the untreated alloy—without electron-beam processing (30–35 μm) in mode 1 ($E_S = 50$ J/cm^2), whereas it rises to 40–45 μm in mode 2 ($E_S = 60$ J/cm^2). The thickness of the intermediate layer extends from 15–17 to 45–50 μm in both cases.

A dip of microhardness is drawn attention to at depths of 10–15 and 25–30 μm in mode 1 and 2, respectively. It is reported on the origination of a volumetric microhardness maximum. Here, a local minimum of microhardness correlates with the penetration depth of the alloyed zone under electron-beam processing. The explanation is that pulsed-periodic electron-beam processing in the modes above causes a far longer time the surface is in the molten state than in electrical explosion alloying. On one hand, the element composition of the surface layer material to be melted

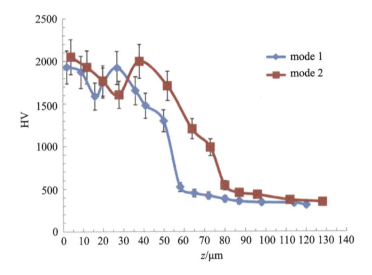

Fig. 8.4 Depthwise distribution of microhardness in a zone alloyed in the electrical explosion of a TiB$_2$ powder portion (250 mg) and irradiated by an electron beam in conditions: 1—$E_S = 50$ J/cm^2, $\tau = 100$ μs, $N = 10$ impulses, $f = 0.3$ s^{-1}; 2—$E_S = 60$ J/cm^2, $\tau = 100$ μs, $N = 10$ impulses, 0.3 s^{-1}

homogenizes, on the other hand, annealing effects still remain in it during crystallizing. As a result, structure of the remelted layer gets refined and microhardness increases. Less significant transformations are observed in a lower part structure of the surface layer under electron-beam processing. Therefore, a microhardness maximum might be referred to as the distribution of residual stresses on the boundary between remelted and solid zones of the surface layer in the alloyed zone under electron-beam processing.

Electron-beam processing of the alloyed surface by a weighted portion of B$_4$C powder. If powder B$_4$C is used as an alloying additive, it is reported, similarly to the case above, on an enhancement of microhardness from 2400 to 4000 HV under electron-beam processing in mode 2. Other outcomes are that a local microhardness minimum appears at a depth of about 20 μm and a thickness of the surface layer rises to 30–40 μm and that of the intermediate layer to 35–40 μm (Fig. 8.5).

Electron-beam processing of the alloyed surface by a weighted portion of SiC powder. In both modes under consideration the subsequent electron-beam processing of the alloyed zone results in a drop of microhardness (to 600 HV on average) in the surface and subsurface layers (Fig. 8.6). Similarly to both cases described above, a volumetric microhardness minimum is detected. Electron-beam processing produces no significant effect on the thickness of the zone to be hardened. A decrease instead of an increment of microhardness as in two prior cases might be attributed to a certain interaction degree between silicon carbide and titanium and its alloys. Additionally, a raise in an electron-beam energy density reduces a number of volumetric microhardness maximums: 2 maximums (Fig. 8.6, curve 1) if irradiated by an electron beam

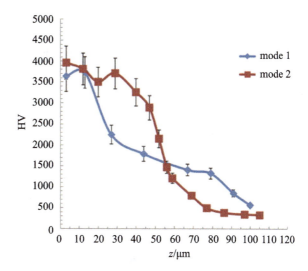

Fig. 8.5 Depthwise distribution of microhardness in a titanium alloy VT6 surface alloyed in the electrical explosion of a B$_4$C powder portion (496 mg) and irradiated by an electron beam in conditions: 1—E_S = 50 J/cm^2, τ = 100 μs, N = 10 impulses, f = 0.3 s^{-1}; 2—E_S = 60 J/cm^2, τ = 100 μs, N = 10 impulses, 0.3 s^{-1}

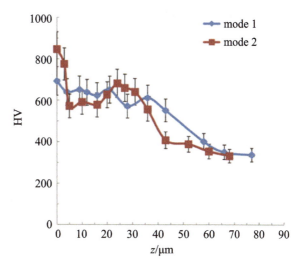

Fig. 8.6 Depthwise distribution of microhardness in a surface alloyed in the electrical explosion of a SiC powder portion and irradiated by an electron beam in conditions: 1—E_S = 50 J/cm^2, τ = 100 μs, N = 10 impulses, f = 0.3 s^{-1}; 2—E_S = 60 J/cm^2, τ = 100 μs, N = 10 impulses, 0.3 s^{-1}

(E_S = 50 J/cm^2); 1 maximum at E_S = 60 J/cm^2 (Fig. 8.6, curve 2). The remotest maximum of microhardness is determined to be in the intermediate layer.

To sum up, the electrical explosion alloying of titanium alloy VT6 by silicon carbide multiplies the microhardness of the surface layer approximately by six times; the total thickness of the hardened layer is about 40 μm. As a result of the subsequent electron-beam processing, the microhardness of the surface layer rises by 3.5 times (as compared to the microhardness of a sample volume), and the hardened subsurface layer is as thick as 80 μm.

Characteristics of the alloyed zone developed under complex processing by different weighted powder portions may depend on a diversity of physical and chemical properties of the materials used. For instance, the microhardness in the surface layers correlates with that of materials used as weighted powder portions: its values for TiB_2, B_4C, and SiC are 3480, 4910, and 3300 HV, respectively [8–14].

8.2 Wear Resistance of Modified Layers

The electrical explosion alloying of titanium alloy VT6 by a weighted powder portion TiB_2 (250 mg) enhances the wear resistance of the surface by 14.5 times (Figs. 8.7, 8.8 and 8.9, Table 8.1) if compared with the initial state. As a result of the subsequent electron-beam processing, wear resistance is heightened by more than 90 times, and a friction coefficient is two times lower.

To conclude, the depthwise distribution of microhardness exhibits two layers. The microhardness of the surface layer is determined to be reasonably high (900, 850, and 1100–1200 HV for weighted powder portions of TiB_2, B_4C, and SiC, respectively). It declines monotonously to the initial level in the intermediate layer. The total thickness of the hardened zone is reported to be several tens of micrometers.

A significant rise of a powder portion weight furthers a multifold increase in microhardness (to 1200 and 2400 HV for weighted portions of powders TiB_2 and B_4C, respectively).

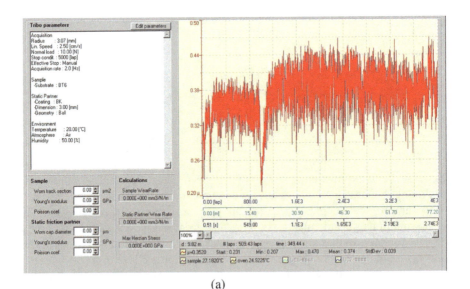

(a)

Fig. 8.7 Friction coefficient vs. length of friction track (**a**) and characteristic profiles of a wear track (**b**)

8.2 Wear Resistance of Modified Layers

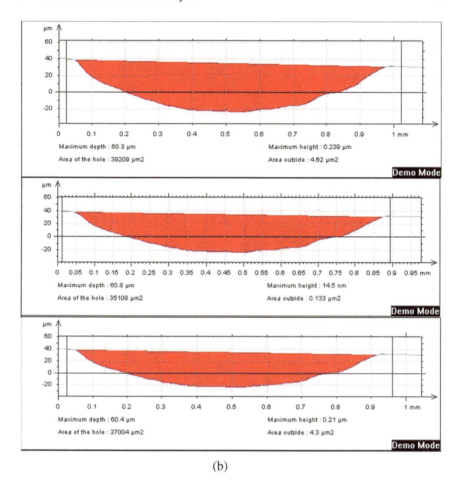

(b)

Fig. 8.7 (continued)

The subsequent electron-beam irradiation using powders TiB_2 and B_4C raises microhardness to 1800 and 4000 HV, respectively. If irradiated by electron beams after the alloying by a weighted powder SiC portion, the microhardness decreases.

A local minimum correlating with a penetration depth into the upper part of the surface is found in the depthwise microhardness pattern after electron-beam processing.

As a result of the electrical explosion alloying by a weighted powder portion TiB_2, the wear resistance enhances by 14 times in dry friction conditions, and a friction coefficient is twice as low. The further electron-beam processing results in a 90 fold increase in wear resistance.

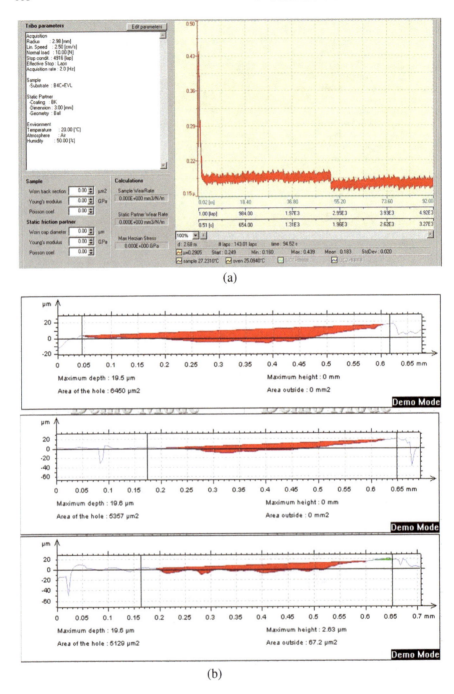

Fig. 8.8 Friction coefficient vs. length of friction track (**a**) and characteristic profiles of a wear track (**b**) after electrical explosion alloying

8.2 Wear Resistance of Modified Layers

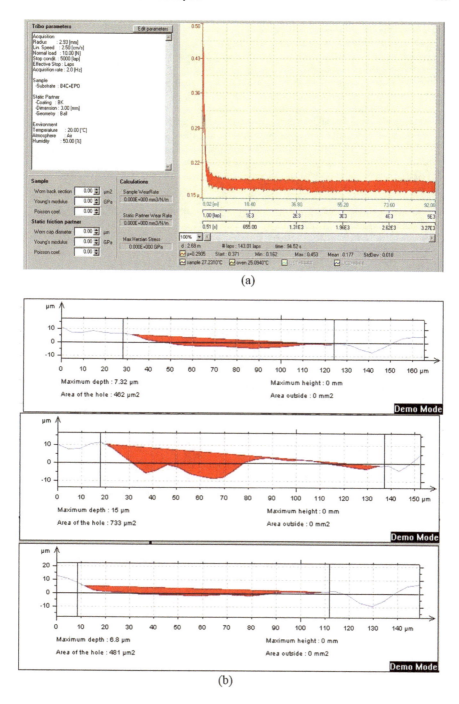

Fig. 8.9 Friction coefficient vs. length of friction track (**a**) and characteristic profiles of a wear track after electrical explosion alloying and electron-beam processing (**b**)

Table 8.1 Tribological characteristics of the processed alloy VT6

State of the material[a]	$W/(\times 10^{-6}$ mm^3/N·m)	$<\mu>$	μ(min)	μ(max)
1	910.7	0.374	0.207	0.47
2	62.2	0.183	0.16	0.439
3	10.3	0.177	0.162	0.453

[a]*Note*: 1—as-cast state; 2—after electrical explosion alloying; 3—after electrical explosion alloying and the subsequent electron-beam processing (60 J/cm^2, 100 μs, 10 impulses, 0.3 s^{-1})

References

1. Cui, W.: A state-of-the-art review on fatigue life prediction methods for metal structures. J. Mar. Sci. Technol. **7**, 43–56 (2002)
2. Stephens, R.I., Fatemi, A., Stephens, R.R., Fuchs, H.O.: Metal Fatigue in Engineering. Wiley Interscience, New York (2000)
3. Schutz, W.: A history of fatigue. Eng. Fract. Mech. **54**, 263–300 (1996)
4. Bhaduri A.: Fatigue. In: Mechanical Properties and Working of Metals and Alloys. Springer Series in Materials Science, vol. 264. Springer, Singapore (2018)
5. Yokobori, T.: Physics of Strength and Plasticity. MIT Press, Boston (1969)
6. Kennedy, A.J.: Processes of Creep and Fatigue in Metals. Oliver & Boyd Ltd., London (1962)
7. Golovin, S.A., Puskar, A.: Microplasticity of a low-carbon steel at low and high loading frequencies. Kov. Mater. **16**, 426–437 (1978)
8. Collacott, R.A.: Mechanical Fault Diagnosis and Condition Monitoring. Chapman & Hall, London, UK (1977)
9. Pisarenko, G.S., Troshchenko, V.T., Strizhalo, V.A., Zinchenko, A.I.: Low-cycle fatigue and cyclic creep of metals. Fatigue Fract. Eng. Mater. Struct. **3**, 305–313 (1980)
10. Krautkrämer, J., Krautkrämer, H.: Ultrasonic Testing of Materials. Springer-Verlag, Berlin Heidelberg (1990)
11. Ivanova, V.S., Goritskii, V.M., Orlov, L.G., Terent'ev, V.F.: Dislocational structure of iron at the tip of a fatigue crack. Strength Mater. **7**, 1312–1317 (1975)
12. Miller, K.J.: Metal fatigue—past, current and future. Proc. Inst. Mech. Eng. Part C J. Mech. Eng. Sci. **205**, 291–304 (1991)
13. Shanyavskiy, A.A., Banov, M.D., Zakharova, T.P.: Principles of physical mesomechanics of nanostructural fatigue of metals. I. Model of subsurface fatigue crack nucleation in VT3–1 titanium alloy. Phys. Mesomech. **13**, 133–142 (2010)
14. Novikov, I.I., Ermishkin, V.A.: On the analysis of stress-strain curves of metals. Izv. Akad. Nauk SSSR. Met. **6**, 142–145 (1995)

Chapter 9
Effect of Electron-Beam Processing on Structure and Phase Composition of Titanium VT1-0 Fractured in Fatigue Tests

The chapter outlines the research outcomes into the effect of various electron-beam processing modes on the structure and phase composition of titanium VT1-0. They have been published in works [1–7].

9.1 Fracture Surface, Structures, and Phase Composition of Fractured Titanium VT1-0 When Fatigued

In this section, the authors focus on a fracture surface and structures of the titanium VT1-0 surface layer destroyed in high-cycle fatigue tests. Samples to be analyzed are in as-delivered state.

9.1.1 Fractography of the Fatigue Fracture Surface

The fatigue fracture is usually a time-depending phenomenon developing in local areas of a deforming material. Once a sample is in critical state, it tends to destruction. It is reported on three typical zones on the fracture surface—the fatigue propagation of a crack separated from a rupture area by a zone of enhanced crack propagation. A characteristic fracture surface of samples destroyed under regular fatigue is shown in Fig. 9.1.

The evolution of deformation processes related to fatigue tests of a material depends on a zone, being total in the zone of fatigue crack propagation and far less significant in the rupture area. In most cases, a fatigue fracture originates on the surface of metallic materials, furthering the intensive plastic deformation of a surface layer as deep as a grain size. Figure 9.2 demonstrates a typical structure of a titanium surface layer during fatigue fracturing. As we can see, the fatigue stressing leads to the development of a thin (1.0–1.5 µm) surface layer (Fig. 9.2b).

Fig. 9.1 Structure of titanium fracture surface

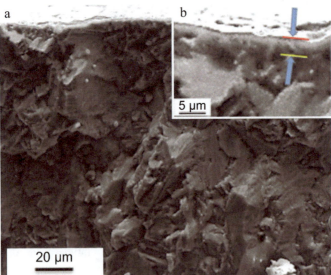

Fig. 9.2 Structure of the fracture surface in a titanium sample after fatigue tests. Arrows (b) indicate a thin surface layer developing in fatigue tests of a sample

9.1 Fracture Surface, Structures …

As a consequence of each load variation, a slight but quite localized plastic deformation occurs at a crack top. It is a polycrystalline structure of titanium that causes the multiple branching of a rupture front in the material. It is reported on the formation of numerous microscopically visible and parallel rupture marks (Fig. 9.3).

Progression marks are considered to represent a decisive characteristic of a fatigue fracture zone in the material. Figure 9.4 shows a typical fracture surface of commercially pure titanium with progression marks. Generally speaking, progression marks are bands of consecutive low and high spots or bands parallel to a crack front with detachment steps surrounded by these hollows. Once a loading cycle has been completed, a crack (rupture) moves forward at a certain distance. Simultaneously, a row of consecutive bands is generated on a fracture surface. Therefore, bands indicate the crack movement, normally one step per loading cycle.

Schmitt-Thomas and Klingele have suggested referring to these bands as progression marks. Progression marks can be perpendicular or almost perpendicular to the

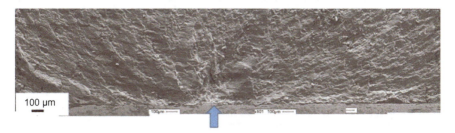

Fig. 9.3 Electron-microscopic view of a titanium structure fractured in fatigue tests; the branching of a material rupture front is apparent. The arrow indicates a center of crack propagation

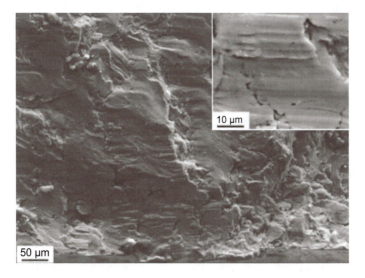

Fig. 9.4 Progression marks developing in titanium as a result of fatigue fracture

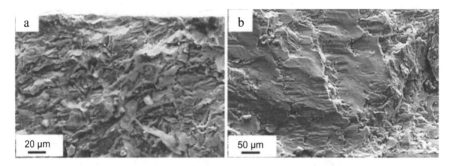

Fig. 9.5 Electron-microscopic view of fatigue fracture titanium surface: (a) structure of the fatigue crack propagation; (b) structure of the rupture area

direction of a propagating crack. They can be continuous and regular (typical for aluminum alloys) with a narrowing gap between them if reduced stress and speeds of crack propagation. They can also be broken and irregular, such as those on fracture surfaces of steels. The space between marks depends on how the material resists the fatigue crack propagation, with all other factors of fatigue tests kept the same: the narrower the space between marks, the higher resistance to the propagation of a crack material has. In our study, the average space between progression marks in fractured titanium samples was assessed to be 3.0 μm.

In most cases, a fracture surface is characterized by a complex structure. As reported, the integrated mechanism of fatigue fracture is typical for metallic materials. In the data (Fig. 9.5) of electron-microscopic views, ductile fracture pits and quasi-spalling facets are found. Ductile fracture pits—detected mainly in the zone of fatigue crack propagation—develop due to the cutting of micro-pores, which have facilitated the fracture of titanium grains (Fig. 9.5a). Quasi-spalling facets are a principal element in the fracture surface of the rupture area (Fig. 9.5b).

To sum up, the research into the fracture surface of titanium VT1-0 carried out employing the SEM methods revealed a thin surface layer, a zone of fatigue fracture and a rupture area. It was demonstrated that a zone of fatigue fracture contains progression marks spaced 3.0 μm apart. According to the study, ductile fracture pits were detected in the zone of fatigue crack propagation; furthermore, quasi-spalling facets are a main structural element in the fracture surface of a rupture area.

9.1.2 Defect Sub-Structure and Phase Composition of the Titanium Surface Layer Fractured When Fatigue Testing

The defect sub-structure of the titanium surface layer fractured in the course of fatigue test was analyzed by the methods of TEM using thin foils. The study has

9.1 Fracture Surface, Structures ...

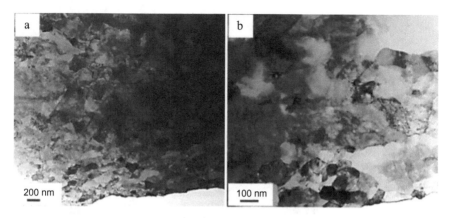

Fig. 9.6 Cross-section structure in titanium sample fractured when fatigue testing

demonstrated that fatigue tests of titanium initiate the development of a thin submicro nano-crystalline surface layer (Fig. 9.6).

A relatively thin (≈ 2 μm) surface layer is characterized by a polycrystalline structure. Grains (sub-grains)—elements of this layer vary in a range from 30 nm to 70 nm (Fig. 9.6b). In addition, a surface layer has a lamellar structure, i.e. grains of the surface layer form sub-layers parallel to the sample surface (Fig. 9.6b). The thickness of a sub-layer is similar to the average grain size. The regular lattice-like sub-structure turns into a stripe-like dislocation sub-structure at a distance below the sample surface (Fig. 9.7a). Layers with a nano-scale fragmented sub-structure are found along the stripe-like sub-structure, at boundaries and grain joints (Fig. 9.7b). The structure of randomly distributed dislocations is detected throughout the stripes. The scalar density of dislocations was measured to be 2.5×10^{10} cm^{-2}.

Cellular and lattice-like structure represents another type of dislocation substructure found in titanium grains fractured under fatigue (Fig. 9.7c). Colonies of micro-twins are detected in joints of such grains (Fig. 9.7d).

The phase composition of a titanium surface layer fractured in fatigue tests was explored using the methods of diffraction electron microscopy via indexing corresponding electron-diffraction micro-patterns and researching dark-field images. Figures 9.8 and 9.9 present the research data into the structure and phase state of a nano-structured surface layer.

Foremost, researchers emphasized a circular nature of the electron-diffraction micro-pattern, suggesting the development of a nano-dimensional grain structure in the surface layer. This is clearly evidenced by a dark-field analysis of the surface structure in the reflex of α-titanium (Fig. 9.9a). Crystallites are anisotropic; a longitudinal axis of crystallites is parallel to the sample surface. The surface layer is a multiphase aggregate. The indexing of an electron-diffraction micro-pattern (Fig. 9.8b), obtained on this layer, has pointed out reflexes of titanium oxides Ti_3O_5 (Fig. 9.8b, reflex 2) and Ti_2O (Fig. 9.8b, reflex 3). Oxide phase particles form layers along titanium grain

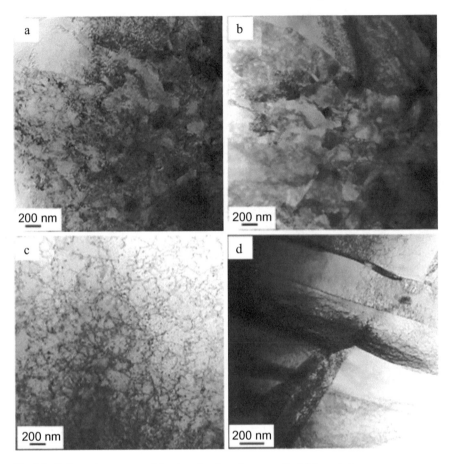

Fig. 9.7 Defect sub-structure of titanium developed as a result of fatigue tests

boundaries and throughout titanium grains, obviously, on dislocations. Oxide phase particles vary in a range from 10 to 15 nm.

Fragmented sub-structure is an area where the ductile fracture of a material is initiated. This process is suggested to evolve in the following way. As deformation proceeds, local spots unable to evolve further (a so-called critical structure) tend to develop in the fragmented structure. The results of comprehensive crystallographic studies demonstrated micro-cracks open along certain boundaries and may stop close to certain joints. In most cases, these boundaries are with a decisive plastic angle of adjacent crystalline areas (more than 15–20 degrees), i.e. the rotation instability as a disclination defect develops in such areas of a strained sample.

Similar results are reported in later research works [10–13] focused on the structure of steel 20X23H18 exposed to fatigue tests. Reconsidering the given above investigation results of titanium VT1-0, we can conclude zones of the material with a critical structure unable to the more significant evolution appear in the surface layer

9.1 Fracture Surface, Structures …

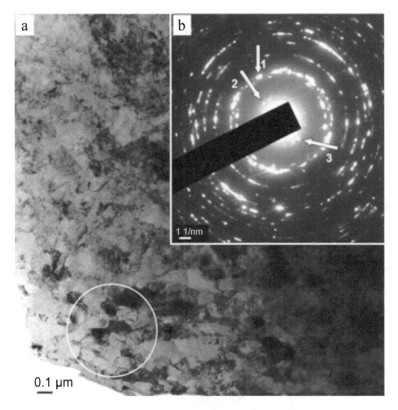

Fig. 9.8 Structure of the sample cross-section fractured during fatigue testing: (a) bright-field view of the surface layer (low edge of the picture complies with the sample surface); (b) electron-diffraction micro-pattern obtained on the oval area (a)

up to 2 μm because of fatigue tests. It is suggested that a structure in the material under consideration is a polycrystalline surface layer with crystallites within a range of 100 nm. The development of this structure might initiate the micro-crack origin, leading, therefore, to the fracture of a whole sample.

To sum up, the research has provided an evidence that titanium VT1-0 tends to fracture in conditions of fatigue tests owing to the formation of a surface layer with a critical structure failing to evolve further, i.e. with the exhausted plasticity (fatigue endurance)—a material with nano-crystalline structure. A thickness of the nano-crystalline surface layer is determined to be ≈ 2 μm; additionally, oxide phase inclusions were detected along the boundaries and throughout nano-dimensional grains of α-titanium.

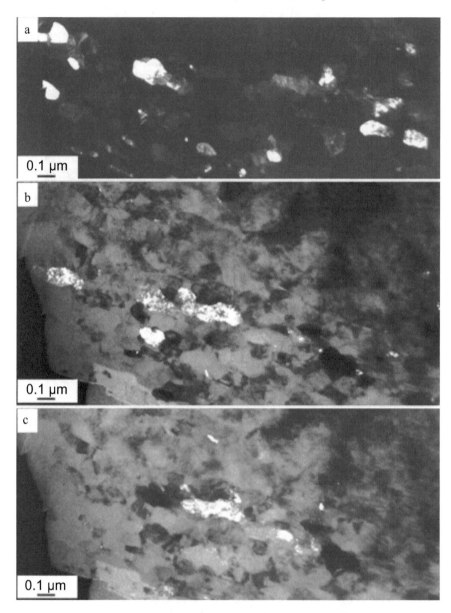

Fig. 9.9 Dark-field views of foil section in Fig. 9.8a. (a) the dark field obtained in [8] α-Ti; (b) the dark field obtained in [9] Ti_3O_5 (reflex 2 [Fig. 9.8b]); (c) the dark field obtained in [8] Ti_2O (reflex 3 [Fig. 9.8b])

9.1.3 Structure of Titanium Fractured in Fatigue Tests

The fatigue failure of metal and alloys is related to the origin and propagation of cracks in the surface layer of a machine part. Therefore, a state of the surface plays a vital role for the fatigue endurance of a material.

This section aims to examine the fatigue fracture regularities of titanium VT1-0 after the high-temperature pre-annealing in air. High-cycle fatigue tests of this sample resulted in its fracture after 233,512 loading cycles.

Figure 9.10 shows a typical fracture surface of titanium exposed to high-temperature annealing in air. From these data it is apparent that a multilayered structure comprising a thin surface up to 2 μm (Fig. 9.10b, the layer is shown by dotted lines) and an intermediate layer up to 100 μm (Fig. 9.10a, the layer is indicated by lines) is a product of fatigue tests.

The surface layer (Fig. 9.10b) is separated by micro-pores and micro-cracks from the intermediate layer. Assumingly, this surface layer is formed due to the prior oxygen saturation of the material when annealed in air.

A characteristic fracture surface of commercially pure titanium with progression marks is given in Fig. 9.11.

From the TEM data of the surface layer (Fig. 9.12), it is seen that the thickness of a layer to be analyzed H is ≈ 110 μm.

As shown in Fig. 9.10c, the fatigue rupture of the titanium sample has a thin surface layer. The structure of this layer explored by the methods of TEM is indicted in Fig. 9.13.

It is apparent that a surface layer (2.8 μm) possesses a sub-micro and nano-crystalline structure. Crystallites, constituting this layer, range from 80 to 120 nm (Fig. 9.13b). Second phase inclusions varying from 12 to 15 nm are detected on the

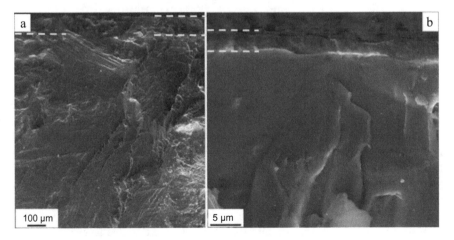

Fig. 9.10 Structure of titanium fracture surface. Dotted lines (a) and (b) indicate the intermediate (a) and surface (b) layers

Fig. 9.11 Progression marks in titanium—a product of fatigue fracture

edges of crystallites (Fig. 9.14). Electron-diffraction micro-patterns obtained on this layer demonstrate reflexes of titanium oxides Ti_2O and TiO_2 (rutile) (Fig. 9.13d, reflexes indicated by arrows).

The layer close to the nano-structured surface is α-titanium with a great number of micro-twins (Fig. 9.15). The highest density of micro-twins (a number of micro-twins per unit of a secant line length) is recorded at the boundary with a nano-structured surface layer; all things considered, it is an evidence of severe stresses developing in this material layer during fatigue testing.

The number of micro-twins reduces gradually below the surface layer (Fig. 9.16a).

From the electron-diffraction micro-patterns obtained on this layer the azimuth migration of reflexes is apparent, indicating the formation of a disoriented structure. The thickness of the layer with a lamellar sub-structure is determined to be $\approx 20\ \mu m$.

9.1 Fracture Surface, Structures … 181

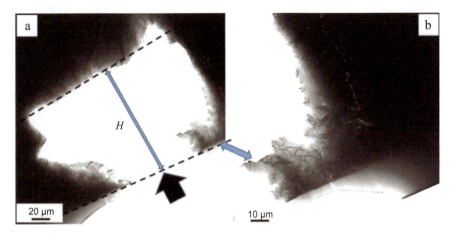

Fig. 9.12 Electron-microscopic view of the titanium surface fractured in fatigue tests. The double arrow (a) indicates a thickness of the analyzed foil layer H, (b) a magnified view of a foil zone (a); the black arrow points at the sample surface

As far from the sample surface a dislocation sub-structure only is detected in α-titanium grains, its typical view is given in Fig. 9.16b.

It is reported on a thin (≈ 2 μm) surface layer with a sub-micro and nano-crystalline structure separated from the main volume of the material by micro-pores and micro-cracks. The layer close to the nano-structured surface layer is determined to be α-titanium with numerous micro-twins. Probably, it evidences severe stresses developing in this material layer during fatigue testing. The thickness of the layer with lamellar structure is measured to be 20–25 μm. The number of micro-twins decreases at a distance from the nano-structured surface layer. A principal element of the defect sub-structure is chaotically located dislocations in α-titanium at a farther distance below the sample surface.

9.1.4 Gradient Structure Developing in Titanium When Fatigued

TEM methods of thin foils in the STEM mode used in the study allowed to establish a gradient structure develops during fatigue testing; its feature is a regular variation of defect sub-system states in the sample (Fig. 9.17). From the data (Figs. 9.17 and 9.18) it is seen a relatively thin (≈ 2 μm) surface layer with a polycrystalline structure forms owing to fatigue tests. Grains (sub-grains)—elements of this layer vary in a range from 30 to 70 nm. The grains constitute layers parallel to the sample surface.

The thickness of the lamellar structure in the surface layer equals to the average grain size. At a distance below the sample surface, a distortion in the lamellar structure with a nano-dimensional grain sub-structure is observed (Figs. 9.17 and 9.19). A band

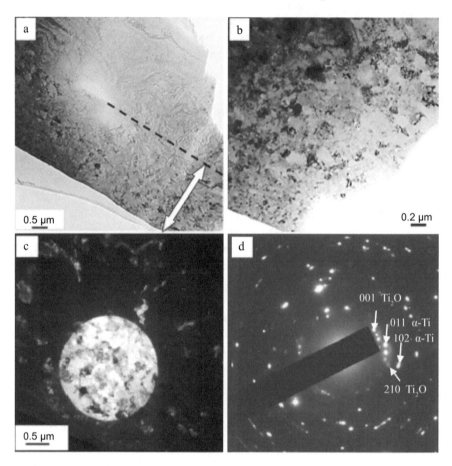

Fig. 9.13 Electron-microscopic view of the titanium surface fractured in fatigue tests. The circle (c) (selective diaphragm) shows a foil section where the data were gathered (d)

sub-structure is reported. Layers with nano-dimensional fragmented sub-structure are detected along the band sub-structure. A dislocation sub-structure in the form of chaotically distributed dislocations is found throughout the bands. The scalar density of dislocations is determined to be 2.5×10^{10} cm^{-2}. The band sub-structure is substituted by a cellular and mesh-like dislocation sub-structure at a distance below the sample surface (Fig. 9.20). Colonies of micro-twins are found in the joints of grain boundaries (Fig. 9.21). At a distance of 30–35 μm below the sample surface, a basic element in the grain volume is a sub-structure of the dislocation chaos (Fig. 9.22).

To analyze the phase composition of the surface layer in titanium VT1-0 fractured during fatigue testing, authors applied methods of electron-diffraction microscopy, e.g. the indexing of electron-diffraction micro-patterns and dark-field images. Figures 9.23 and 9.24 provide the research data on the structure-phase state of the nano-structured layer.

9.1 Fracture Surface, Structures …

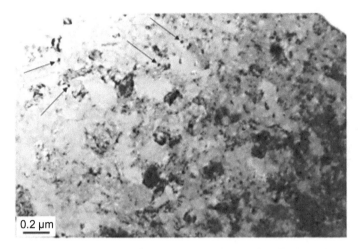

Fig. 9.14 Electron-microscopic view of the titanium surface fractured when fatigue testing. Arrows show particles of the oxide phase

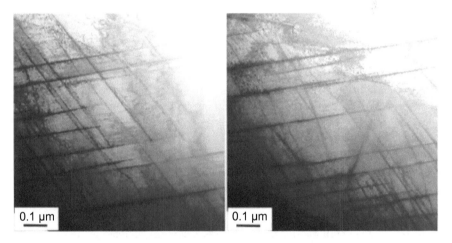

Fig. 9.15 Electron-microscopic view of a titanium structure fractured when high-cycle fatigue testing

A principal feature is a circular character of the electron-diffraction micro-pattern, indicating the development of a nano-dimensional grain structure in the surface layer. Unambiguous evidence is observed in the dark-field analysis data of the surface layer structure in α-titanium reflexes (Fig. 9.24a). The crystallites are anisotropic; a longitudinal axis of crystallites is parallel to the sample surface.

The surface layer represents a multiphase aggregate. The indexing of an electron-diffraction micro-pattern obtained on this layer has revealed reflexes of titanium oxides Ti_3O_5 (Fig. 9.24b) and Ti_2O (Fig. 9.24c). Oxide phase particles constitute

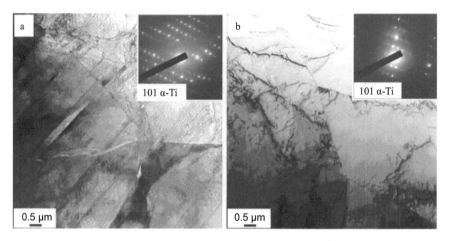

Fig. 9.16 Electron-microscopic view of a titanium structure fractured during high-cycle fatigue testing. (a) layer close to the surface nano-crystalline layer; (b) layer at a distance of 45–50 μm below the sample surface

layers along the boundaries of titanium grains and throughout the titanium grains, obviously, on dislocations. Particles of the oxide phase vary in a range from 10 to 15 nm.

Importantly, oxide phase particles are detected mainly in the nano-crystalline layer. At a distance of 4–5 μm below the sample surface, reflexes of the oxide phase were unfound using the methods of electron micro-diffraction microscopy.

Several conclusions have been made when analyzing the dislocation sub-structure vs. distance relation.

First, a scalar density of dislocations increases in the volume of α-titanium grains. For instance, using the secant method, a scalar density of dislocations was determined to be 1×10^{10} cm^{-2} at a distance of 120 μm below the fracture surface, whereas it was 1.5×10^{10} cm^{-2} at a distance of 60 μm and 2.2×10^{10} cm^{-2} at a distance of 35 μm.

Second, a change in the dislocation type is recorded, i.e. chaotically distributed dislocations in titanium grains dominate at a distance of 120 μm below the fracture surface. Besides a structure of the dislocation chaos, there are dislocation agglomerates at a distance of 80 μm, and a prevailing dislocation sub-structure in titanium grains is dislocation meshes at distances of 60 μm and 35 μm.

Additionally, a phenomenon of micro-twinning is observed, and a number of micro-twins in a grain increases as close to the fracture surface. It is obvious that the developing micro-twinning structure is a relaxation mechanism of inner stresses arising in the material during fatigue testing. The importance of micro-twinning has been comprehensively studied on austenite steels [14].

Another crucial fact is the development of a fragmented sub-structure along the boundaries of titanium grains at a distance of 60 μm and deeper below the fracture surface. A fragmented sub-structure is important for a variety of materials because

9.1 Fracture Surface, Structures …

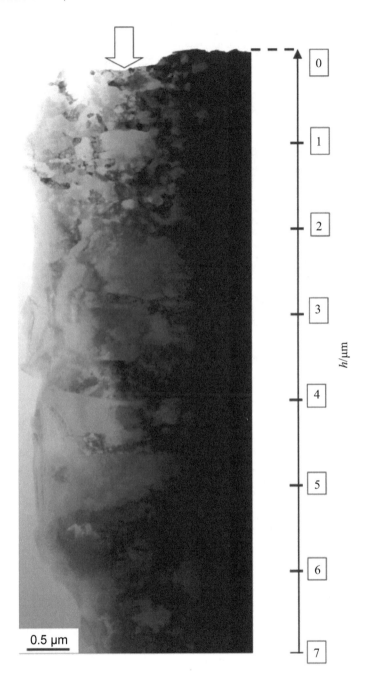

Fig. 9.17 Cross-section structure of titanium fractured in fatigue tests

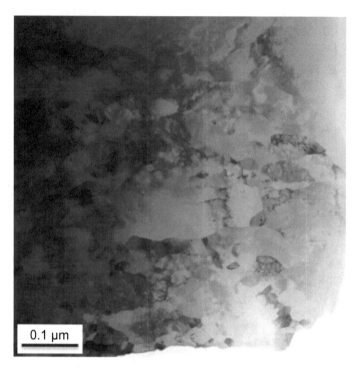

Fig. 9.18 Electron-microscopic view of a layer close to the sample surface

it is an ultimate element in a chain of sub-structure transformations, and further fragments can reduce in size only [15–17]. Subsequently, micro-cracks appear on the boundaries of fragments. A distinctive feature of the fragmented sub-structure is suggested to be an ability to accumulate on its boundaries atoms of interstitial elements, e.g. hydrogen and oxygen. This process might also facilitate the cracking.

Moreover, far-reaching (inner) stresses become more intensive as close to the fracture surface. Far-reaching (inner) fields of stresses in thin foils are reported when bend extinction contours start appearing. The development of far-reaching stress fields is related to the polarization of a dislocation sub-structure and the development of dislocation and disclination sub-structures, the incompatible deformation of adjacent grains, crystallites of various phases, and micro-cracks.

Finally, a grain and sub-grain structure is recorded in the fractured area. The crystallites vary from 120 to 500 μm. Numerous micro-cracks are detected along the boundaries of crystallites; they might be a result of material deformation and associated with the fabrication of thin foils because a high level of stress fields in the material furthers the quick thinning of a sample.

The study carried out by the methods of electron-diffraction microscopy has revealed nano-dimensional (10–15 nm) inclusions of titanium oxide in the fractured

9.1 Fracture Surface, Structures …

Fig. 9.19 Electron-microscopic view of a layer at a depth of 2–4 μm

area. Assumingly, particles of the oxide phase might form during the sample degradation during fatigue testing, and after it, representing a result of the free surface oxidation.

Therefore, the defect sub-structure of grains close to the fatigue fracture titanium VT1-0 surface behaves similarly to the evolution of a defect sub-structure typical for the increasing plastic deformation in metals and alloys.

9.2 Fracture Surface, Structures, and Phase Composition of Commercially Pure Titanium Disintegrated When Fatigued After Electron-Beam Processing

9.2.1 Structure of Titanium Irradiated by a Pulsed Electron Beam

Figure 9.25 provides a characteristic electron-microscopic view of a foil produced via ion thinning, which allows the investigation of the surface layer structure developed

Fig. 9.20 Electron-microscopic view of a cellular mesh-like dislocation substructure at a depth of 5–15 μm

when the sample surface is irradiated by a pulsed electron beam in relation to a distance from the surface being modified.

Figure 9.26 provides typical views of titanium VT1-0 sample surface formed after the irradiation by an intensive pulsed electron beam. The noteworthy feature here is that the irradiation of a titanium sample by an intensive pulsed electron beam in the mode of a melting surface layer results in the development of a uniform structure without micro-pores and micro-craters (Fig. 9.26a), registered usually in electron-beam processing of multiphase materials. It is apparent from the data in Fig. 9.26 that a polycrystalline structure forms in the surface layer owing to the high-speed cooling of a sample from the melting temperature. A lamellar sub-structure is detected in the volume of grains in the surface layer (Fig. 9.26b–d).

For the purpose of a comprehensive analysis of the titanium surface modified by an intensive electron-beam methods of transmission electron microscopy were used in the study. A developing structure was investigated in areas presented in Fig. 9.25, i.e. in layers produced at various temperatures. It has been pointed out that the material of interest is a polycrystalline aggregate irrespectively to the zone and area being investigated; however, the sub-structure of its grains depends significantly on a zone being studied. A lamellar structure (Fig. 9.27, lamellae shown by dark arrows) is seen in the volume of grains in zone 1 (layer I in Fig. 9.25). Principally, the lamellae appear

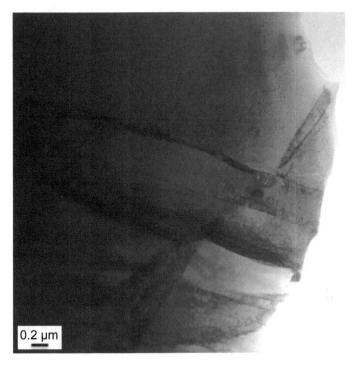

Fig. 9.21 Electron-microscopic view of micro-twins at grain boundaries and their joints. A layer at a depth of ≈ 20 μm

at grain boundaries, there is a space between them; less frequently they demonstrate different physical configuration.

The study has revealed a sub-micro crystalline structure to develop in layer II (Fig. 9.25); its typical view is provided in Fig. 9.28a. Sub-grains vary in a range from 1 to 2 μm.

The structure of a heat impact layer (layer III, Fig. 9.25) depends significantly on the distance to the surface irradiated by an electron beam. For instance, a layer adjacent to layer II (zone 3 (Fig. 9.25)) has a lamellar structure (Fig. 9.28b); a layer at a depth of ≈ 50 μm demonstrates a grain–sub-grain structure shown in Fig. 9.28 c, d. A polycrystalline α-Ti-based structure is recorded at a depth of 180–200 μm (as suggested, a structure of the initial state). A dislocation sub-structure in a form of chaotically distributed dislocations (Fig. 9.29a) or dislocation globules is detected in the volume of titanium grains (Fig. 9.29b). A scalar density of dislocations is determined to be ≈ 1.3×10^{10} cm^{-2}. An average scalar density of dislocations in fractured titanium is two times higher, 2.5×10^{10} cm^{-2}.

To sum up, studies carried out using SEM and TEM in a diffraction mode have established a multilayered state to develop in the surface layer of titanium VT1-0 irradiated by an intensive sub-millisecond low energy pulsed electron beam, the structure of this states depends on the distance to the irradiated surface.

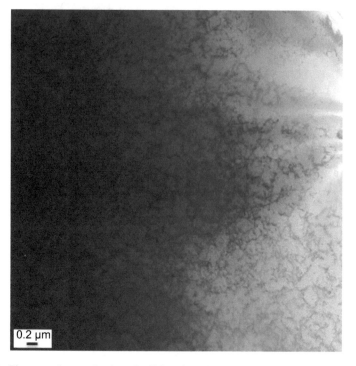

Fig. 9.22 Electron-microscopic view of a dislocation sub-structure in titanium grains at a depth of 30–35 μm

9.2.2 Fracture Surface of Titanium Irradiated by a Pulsed Electron Beam

Fatigue tests of titanium VT1-0 have disclosed a 2.2-time increase of its fatigue endurance on average due to the irradiation by an intensive sub-millisecond pulsed electron beam, an energy density of an electron-beam set 25 J/cm^2. Here, a 3.2-time increase of the fatigue endurance has been recorded in some samples. A raise of an electron-beam energy density up to 30 J/cm^2 has reduced the titanium fatigue endurance. The latter characteristic of titanium samples irradiated at the given electron-beam energy density is reported to be ≈ 40% higher than that of untreated samples.

To begin with, we focus on a structure state of fractured titanium samples irradiated by an electron beam with an energy density of 30 J/cm^2. SEM methods were used to explore the fracture surface structure of commercially pure titanium VT1-0. Figure 9.30 presents a typical view of the fatigue fracture surface in a sample pre-irradiated by an intensive pulsed electron beam (30 J/cm^2). A multilevel character of the rupture is seen in the data. At a macro-level (the fracture surface of a sample on

9.2 Fracture Surface …

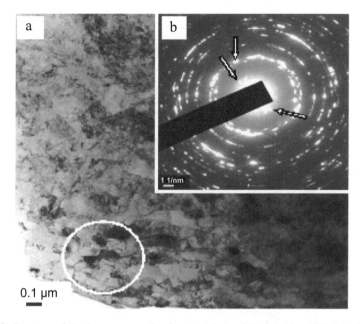

Fig. 9.23 Structure of titanium cross-section fractured during fatigue testing. (a) bright-field view of the surface layer (a bottom line of the figure complies with the sample surface); (b) electron-diffraction micro-pattern obtained on the oval area (a)

the whole) the rupture comprises several blocks with the boundaries (dotted lines in Fig. 9.30) at an angle of $\approx 45°$ to the sample surface.

A surface layer with a thickness up to 100 μm (Fig. 9.31) can be considered as a meso-level of a rupture structure. Micro-pores with a variety of forms and dimensions represent a feature of this layer (Fig. 9.31, micro-pores are shown by arrows).

The surface layer can be conditionally divided into two sub-layers (Fig. 9.32). Further, these sub-layers are referred to as a surface layer (layer 1 in Fig. 9.32) with a thickness of 20–25 μm, and an intermediate layer (50–55 μm) (layer 2 in Fig. 9.32), which is close to the main volume of a sample (layer 3 in Fig. 9.32). Sub-layer 1.1 (≈ 10 μm), which is adjacent to the irradiated surface, can be further singled out in the surface sub-layer, in this layer micro-pores form lines along the boundary with sub-layer 1.2 (Fig. 9.32).

A conclusion can be drawn in the irradiation mode above (30 J/cm^2, 150 μs) a thickness of the surface layer is similar to the thickness of a two-phase layer h_1 (liquid + solid phases), which is produced on the titanium surface irradiated by an intensive electron beam. The thickness of sub-layer 1.1 is comparable to that of a one-phase layer h_2 (liquid phase). Therefore, the surface layer is thought to be a product of the melting titanium surface by an intensive pulsed electron beam.

Each load variation leads to the considerable but quite localized plastic deformation at a crack top. It is suggested that a polycrystalline sample structure (grain structure of titanium) is the reason for the multiple branching of a destruction front

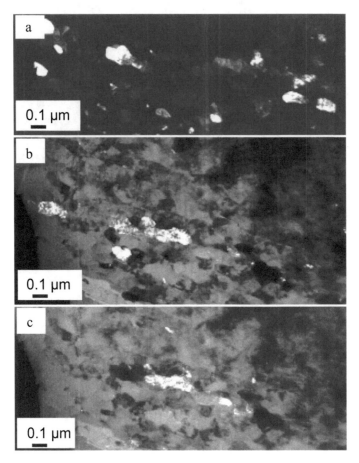

Fig. 9.24 Dark-field views of a foil section shown in Fig. 9.23. (a) dark field obtained in [8] α-Ti (reflex 1 [Fig. 9.23b]); (b) dark field obtained in [9] Ti_3O_5 (reflex 2 [Fig. 9.23b]); (c) dark field obtained in [8] Ti_2O (reflex 3 [Fig. 9.23b])

in the material. A majority of microscopically detectable parallel fracture marks are seen in the data of Fig. 9.33.

Progression marks are considered to be a characteristic of a micro-level in the fatigue fracture zone of commercially pure titanium VT1-0. A typical fracture surface of commercially pure titanium is presented in Fig. 9.34.

The fracture surface tends to have a complex structure because the fatigue fracture has a complex nature. From the data (Fig. 9.35) ductile fracture pits and quasi-spalling facets are seen. The pits are a dominating element in the fracture surface and develop due to the cutting of micro-pores, which have facilitated the fracture of titanium grains.

Therefore, the irradiation product of titanium samples by a pulsed electron beam ($E_S = 30$ J/cm^2) represents a multilayered structure seen in fracture patterns of

9.2 Fracture Surface … 193

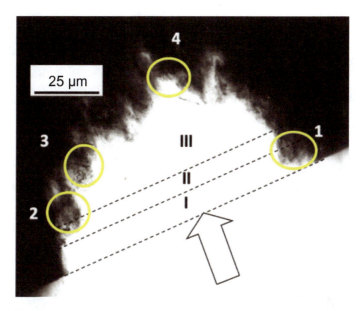

Fig. 9.25 Electron-microscopic view of a foil made via ion thinning of commercially pure titanium. I—one-phase layer (a layer thickness calculated according to a temperature field is as much as 10.4 μm); II—two-phase layer (7.7 μm); III—thermal impact layer. The arrow indicates the surface being irradiated by an electron beam. Ovals show areas of a comprehensive structural electron-microscopic analysis, reprinted from ref. [5] (Copyright 2017, with permission from AIP Publishing)

the material after high-cycle fatigue tests. The studies have pointed out that in the surface layer developed under electron-beam processing, there are micro-pores located parallel to the sample surface. An enormous number of micro-pores in the surface layer might be a key reason for a rather low increase of the fatigue resource in the material of interest.

Essential transformations are observed in the surface layer structure owing to the irradiation of titanium VT1-0 by an electron beam ($E_S = 25$ J/cm^2). The first consequence is the refinement of grains $D = 8.3$ μm on average (Fig. 9.36b, c) (commercially pure titanium VT1-0 in as-delivered state is a polycrystalline material, its average grain is $D = 25$ μm (Fig. 9.36a)).

Second, a lamellar sub-structure is detected in the volume of grains (Fig. 9.36b–d). It evidences the development of additional sub-micro and nano-dimensional structural levels in the surface owing to the electron-beam irradiation.

Figure 9.37 presents a typical view of a fracture surface in commercially pure titanium pre-irradiated by an electron beam ($E_S = 25$ J/cm^2). From the data it is apparent, that the fatigue fracture of the modified sample has a multilayered structure (Fig. 9.37b), i.e. a surface (layer 1), border-line (layer 2), intermediate (layer 3), and a main volume of the material (4).

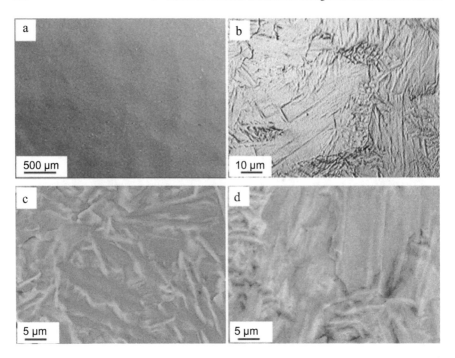

Fig. 9.26 Structure of the irradiated titanium surface formed in electron-beam processing. (a)(c)(d) data obtained by the SEM methods; (b) data of optical microscopy

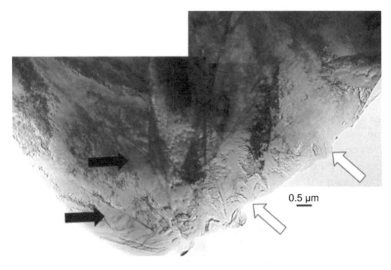

Fig. 9.27 Electron-microscopic view of a structure developing in layer I (zone 1 in Fig. 9.25); light arrows indicate the irradiated surface; dark arrows α-Ti lamellae

9.2 Fracture Surface …

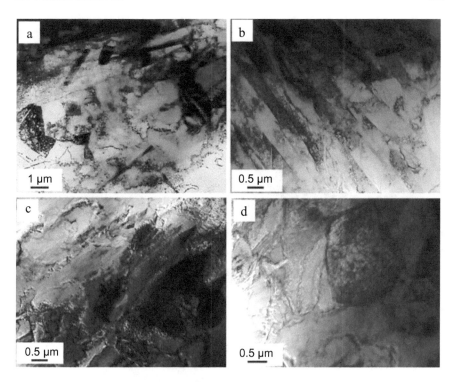

Fig. 9.28 Electron-microscopic view of a structure developing in zone 2 (a); zone 3 (b), and zone 4 (c, d), marked by ovals in Fig. 9.25, reprinted from ref. [5] (Copyright 2017, with permission from AIP Publishing)

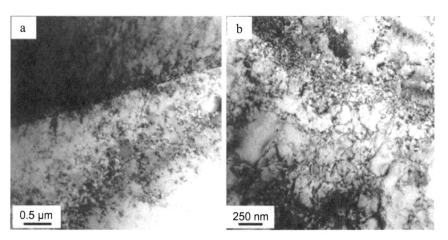

Fig. 9.29 Electron-microscopic view of a dislocation sub-structure at a depth of 180–200 μm in titanium irradiated by an electron beam

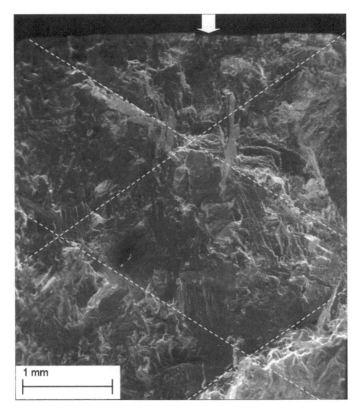

Fig. 9.30 Fracture surface structure of material processed by an electron beam ($E_S = 30$ J/cm^2) before fatigue tests. The arrow indicates the irradiated surface

The fracture structure differs in the singled out layers. The surface demonstrates the smooth fracture; SEM failed to reveal the structure of this layer (Fig. 9.37c). The thickness of this layer varies from 10 to 15 μm in the irradiation mode ($E_S = 25$ J/cm^2). The surface layer can be further divided into two sub-layers (Fig. 9.37c). The thickness of the sub-layer, which forms the irradiated surface, 2–4 μm, is comparable to a thickness of a one-phase layer h_2 (liquid phase). Therefore, the surface (Fig. 9.37c) is a product of the melting and high-speed re-crystallization of commercially pure titanium irradiated by an intensive pulsed electron beam.

The fracture in the border-line layer (25–35 μm) is stream-shaped (Fig. 9.37b, c). Fracture facets in this layer range from 0.4 to 0.6 μm.

The intermediate layer (Fig. 9.37b, layer 3) displays a rougher (if compared to the border-line layer) fracture structure. A distinctive feature of this layer is a significant number of micro-cracks and micro-pores; the layer varies from 0.1 to 0.3 μm (Fig. 9.37d, micro-pores shown by arrows).

To conclude, this section summarizes the data of research into the titanium fracture surface irradiated by a high-intensity pulsed electron beam. The irradiation mode

9.2 Fracture Surface ...

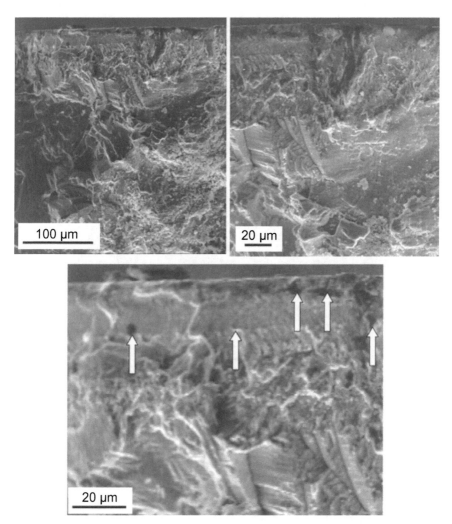

Fig. 9.31 Fracture surface structure of the material processed by an electron-beam ($E_S = 30$ J/cm^2) before fatigue tests

(16 keV, 25 J/cm^2, 150 μs, 3 impulses, 0.3 s^{-1}) was determined to enhance more than twice the fatigue endurance of the material under study. The irradiation of titanium in the mode above causes the refinement of a grain structure and the formation of an inner-grain structure, i.e. additional structural sub-micro and nano-dimensional elements. Fracture patters of the titanium fracture surface irradiated by an intensive pulsed electron beam were analyzed. The data of research pointed out the multilayered structure, which is seen in fracture patterns of the material fractured in high-cycle fatigue tests. It was assumed a lamellar structure of the surface layer represents one of the reasons the fatigue endurance of the material in focus increases significantly.

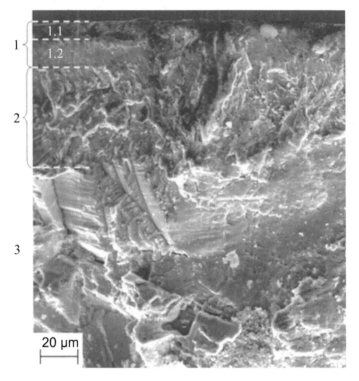

Fig. 9.32 Fracture surface structure of the material processed by an electron-beam ($E_S = 30$ J/cm^2) before fatigue tests

9.2.3 Structure Developed in Fatigue Tests of Samples Irradiated by a Pulsed Electron Beam

The study on titanium VT1-0 has shown an ≈ 40% increase of its fatigue endurance on average as a result of the irradiation by an intensive sub-millisecond pulsed electron beam (16 keV, 30 J/cm^2, 150 μs, 3 impulses, 0.3 s^{-1}) in comparison with untreated samples.

Figure 9.38 demonstrates a typical fatigue fracture in a sample processed by an intensive pulsed electron beam.

To analyze the distribution of temperature in titanium processed by an electron beam, the theoretical evaluation has been carried out.

To determine a temperature field in a certain range of a beam energy density, it is necessary to solve an equation of thermal conductivity. A one-dimensional case of heating and cooling the plate with a thickness d is considered. A system of coordinates is selected that x axis is directed depthwise in a sample. A heat flow is set at $x = 0$, and there is no heat exchange on the rear side of a plate. A thermal conductivity equation is written in coordinates as follows:

9.2 Fracture Surface …

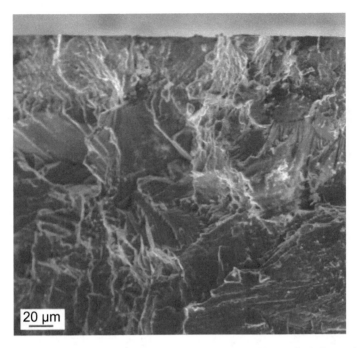

Fig. 9.33 Fracture surface structure in a titanium sample irradiated by a pulsed electron beam ($E_S = 30$ J/cm^2)

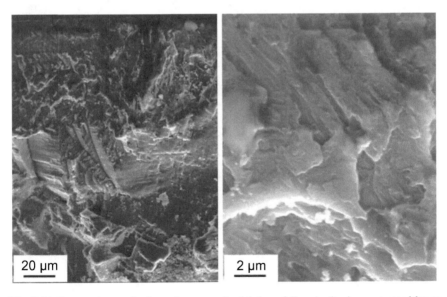

Fig. 9.34 Progression marks formed as a result of fatigue failure in titanium processed by an electron beam

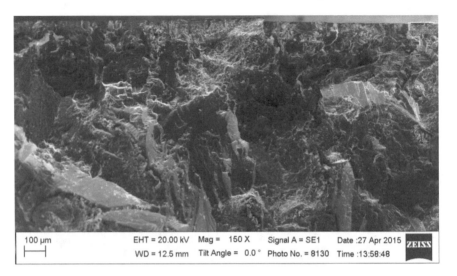

Fig. 9.35 Electron-microscopic view of a structure developing as a result of fatigue fracture in titanium irradiated by an intensive pulsed electron beam

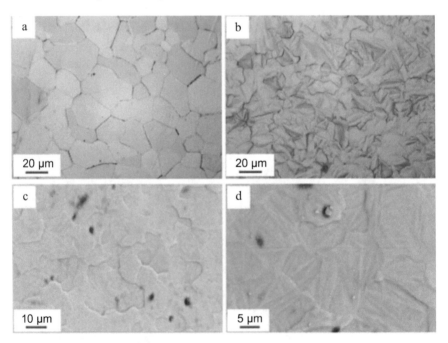

Fig. 9.36 Structure of untreated sample (a); b–d—irradiated by an intensive electron beam ($E_S = 25$ J/cm^2)

9.2 Fracture Surface … 201

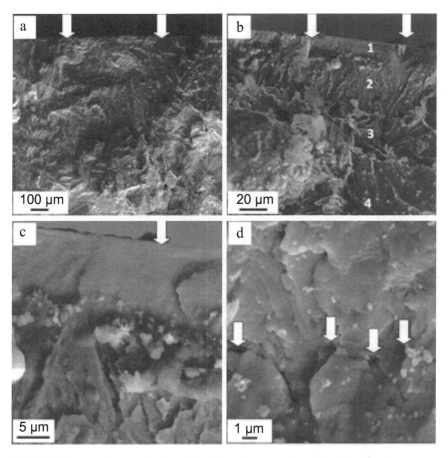

Fig. 9.37 Electron-microscopic view of the fatigue fracture surface in irradiated titanium; arrows (a–c) show the irradiated surface; (d) micro-pores on the boundary between an electron-beam melted layer and a main volume of the material; numbers (b) indicate layers: (1) surface, (2) border-line, (3) intermediate, and (4) main volume of the material

$$C\rho\frac{\partial T}{\partial t} = \lambda\frac{\partial^2 T}{\partial x^2} \qquad (9.1)$$

here C—heat absorption capacity; ρ—density; λ—thermal conductivity of the material.

Boundary (second-type) conditions (Neumann condition) for the pulsed processing by an electron beam are specified:

$$-\lambda\frac{\partial T}{\partial x} = q(t) \qquad (9.2)$$

where a heat flow from the sample surface into its depth is stated as follows:

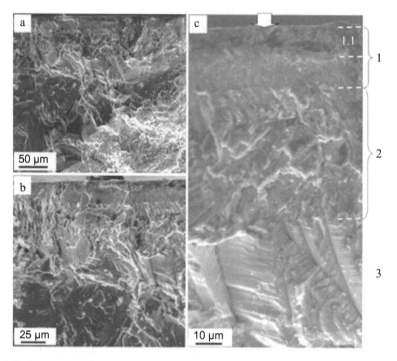

Fig. 9.38 Fracture surface of titanium structure irradiated by an electron beam ($E_S = 30$ J/cm^2) and fatigued. The arrow (c) indicates the irradiated surface. SEM

$$q(t) = \begin{cases} q_0, 0 < t \le t_0 \\ 0, t > t_0 \end{cases} \quad (9.3)$$

here q_0—the mean heat flow over time t_0. There is no heat exchange on the right side of a plate with a thickness d. The input temperature throughout the sample is specified by a value T_0:

$$T(0, x) = T_0, 0 \le x \le d \quad (9.4)$$

The finite difference method is used to solve Eqs. (9.1–9.4). Broadly speaking, the search domain for solutions within the finite difference method is substituted by the seeking for a finite number of points—nodes. In each node the derivatives are replaced by their finite difference approximations, substituted further into a differential equation, as well into initial and boundary conditions. As a result, a differential equation is reduced to the finding of a solution to the system of algebraic equations dependent on values of an indeterminate function in the nodes of a grid.

A number of grid nodes over the plate thickness is specified by a value n, i.e. a step along the axis ox is set $h = d/(n - 1)$. If a total time of the process under study is specified by τ_k and a number of steps over the time—m, one step is written as $\tau =$

9.2 Fracture Surface ...

$\tau_k/(m-1)$. As a consequence, a search domain for a solution is replaced by a grid, the nodes of which have integer coordinates (i, j), $i = 1, \cdots, n; j = 1, \cdots, m$.

An approximation of a thermal conductivity differential Eq. (9.1) with the first time degree and the second space degree is used; moreover, an implicit scheme was selected:

$$\left(\frac{\partial T}{\partial t}\right)_{i,j} = \frac{T_i^{j+1} - T_i^j}{\tau} \tag{9.5}$$

$$\left(\lambda \frac{\partial^2 T}{\partial x^2}\right)_{i,j} = \frac{\lambda}{h^2}(T_{i-1}^{j+1} - 2T_i^{j+1} + T_{i+1}^{j+1}) \tag{9.6}$$

Using the four-point difference scheme—three points are taken on a new time layer and one—on the old one, it is written:

$$\begin{cases} T_i^{j+1} = T_i^j + \frac{\lambda \tau}{\rho C h^2}(T_{i+1}^{j+1} - 2T_i^{j+1} + T_{i-1}^{j+1}), & i = 2, \ldots, n-1; \; j = 1, \ldots, m-1 \\ T_i^1 = T_0; \; i = 1, \ldots n; \\ T_1^{j+1} = T_2^{j+1} + \frac{h \cdot q}{\lambda}; \; j = 1, \ldots, m-1; \\ T_n^{j+1} = T_{n-1}^{j+1}; \; j = 1, \ldots, m-1. \end{cases} \tag{9.7}$$

In addition, an implicitly developed difference scheme (9.7) is absolutely stable.

Thus, a temperature field on a new time layer is represented implicitly, i.e. a solution to the system of algebraic equations is to be found in each internal node:

$$A_i \cdot T_{i+1}^{j+1} - B_i \cdot T_i^{j+1} + C_i \cdot T_{i-1}^{j+1} = F_i \tag{9.8}$$

where

$$A_i = C_i = \frac{\lambda}{h^2}, \; B_i = \frac{2\lambda}{h^2} + \frac{\rho \cdot C}{\tau}, \; F_i = -\frac{\rho \cdot C}{\tau}T_i^j, i = 2, \ldots, n-1.$$

In start and end points of the computational domain, i.e. for $i = 1$ and for $i = n$, a temperature is determined according to boundary conditions on left and right sides of a plate, respectively.

A solution to a system of Eq. (9.8) is to be found on each time layer. For this purpose, the sweep method—a modification of the Gauss method for the three-diagonal matrix (9.8) is usually used. In this case, the predominance of diagonal elements (9.8) making the sweep method stable vs. a rounding error is to be taken into account. For instance, if, $|B_i| \geqslant |A_i| + |C_i|$ at least for one value of I there is a strict inequality; and a system (9.8) has a unique solution.

The melting and evaporation of the material were taken into consideration because of the following physical laws [18]. Once a melting temperature T_{mel} reached a point (j, i), its temperature was recorded on assumption to be equal to the melting temperature, and all the supplied heat:

$$p = \frac{\left(T_i^j - T_{n\pi}\right)C}{q_{n\pi}} \quad (9.9)$$

where T_i^j—the temperature in point (j, i) was assumed to be used for the melting of a sample; $p = 1$ refers to the transition of a substance into a liquid state. The crystallization process was modeled in the same way, only a value p decreased from 1 to 0. The evaporation and condensation process was modeled in the same way:

$$p = \frac{\left(T_i^j - T_{kun}\right)C}{q_{kun}} \quad (9.10)$$

Here $p = 1$ represents the transition of a substance into vapor. A decrease in p from 1 to 0 corresponds to the vapor condensation.

The evaluation was performed for an intensive electron beam with an energy density varying in a range from 2 J/cm² to 30 J/cm², pulses lasted 30, 50, and 150 μs. A thickness of the surface layer for heat calculations d = 0.5 × 10⁻³ m, an observation time—600 μs. Numerically the problem was solved for thermophysical parameters of titanium given in reference materials.

Therefore, the numerical solution to the problem of a temperature field, which develops in the surface layer of titanium irradiated by an intensive electron beam, allowed suggesting the reasons for the formation of a lamellar structure detected when exploring ruptures given in Fig. 9.36.

The findings indicate that the irradiation of titanium by an electron beam in specified parameters leads to the melting of a surface (25.2 μm) (Fig. 9.39).

This layer splits into two sub-layers: a one-phase (liquid) surface (16.6 μm) (Fig. 9.39, layer 1) and a two-phase (liquid + solid state) intermediate layer (8.6 μm) (Fig. 9.39, layer 2). The first layer exists for 128.5 μs; a period from the melting of titanium till the completion of crystallization is 244.2 μs.

Therefore, the results of heat calculations show that the electron-beam processing of titanium samples (30 J/cm², 150 μs) brings about the development of a rather thin surface layer (≈ 25 μm) owing to the high-speed crystallization of the melt. Comparing the results, we assume that the surface layer 1 (Fig. 9.40) is a product of the melting and high-speed crystallization, whereas layer 2 (Fig. 9.40) is a heat impact layer. Furthermore, sub-layer 1.1 forms due to the crystallization of a one-phase (liquid) state; and sub-layer 1.2 because of the crystallization of a two-phase (liquid phase + solid phase) state.

TEM methods were utilized to explore the phase composition and defect substructure of the surface layer in titanium samples pre-irradiated by an electron beam

9.2 Fracture Surface …

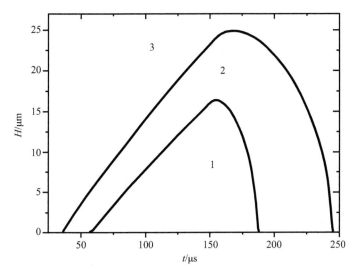

Fig. 9.39 Thickness of one-phase layer 1 (liquid) and two-phase layer 2 (solid phase + liquid) vs. observation time when processing the titanium surface by an electron beam (energy density 30 J/cm^2); zone 3—zone of the solid phase location

and fractured during fatigue testing. The most obvious finding to emerge from this study is that a multilayered structure develops in a fractured sample. Assessing the morphology, a thin surface layer (≈ 5 μm) is reported, a sub-grain structure (shown in Fig. 9.40) is detected in the volume of its grains. Sub-grains represent globules varying in a range from 500 to 700 nm.

A dislocation sub-structure in a form of chaotically distributed dislocations is found in the volume of grains. At a distance below the irradiated surface, a lamellar sub-structure (Fig. 9.41) is detected in the volume of grains in a layer with a thickness of 20–25 μm. A diffraction microanalysis has revealed these lamellae are α-titanium (Fig. 9.42).

A state of the lamellar structure is closely attributed to the distance from the surface irradiated by an electron beam, i.e. a complex sub-grain—lamellar structure (Fig. 9.41a) is replaced by a proper lamellar structure (Figs. 9.41b, 9.42) [21].

The surface with a lamellar sub-structure contacts a layer with a dislocation sub-structure in the volume of its grains, e.g. chaotically distributed dislocations, dislocation globules, and meshes; a scalar density of dislocations is as much as 1.2×10^{10} cm^{-2} (Fig. 9.43). Interestingly, a similar sub-structure is found in fractured titanium samples not pre-irradiated by an intensive electron beam.

Comparing the results of heat calculations and data of electron-diffraction microscopy, a conclusion may be drawn that the lamellar surface layer is a product of the high-speed crystallization of titanium under electron-beam processing.

Therefore, the fatigue tests of titanium irradiated by a sub-millisecond intensive pulsed electron beam ($E_S = 30$ J/cm^2) pointed out an approximately 40% increase of its fatigue endurance in contrast to untreated samples.

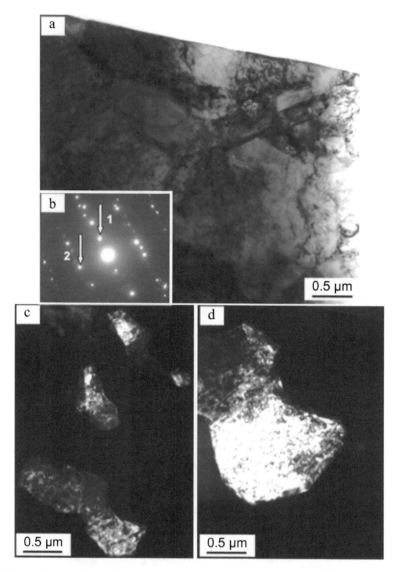

Fig. 9.40 Electron-microscopic view of the irradiated titanium structure fractured when fatigue testing. (a) bright field; (b) electron-diffraction micro-pattern; (c)(d) dark fields obtained in reflexes [19] α-Ti (c) and [20] α-Ti (d). The arrows (b) indicate reflexes, where a dark field was obtained (c)—reflex 1 and a dark field (d)—reflex 2

Fig. 9.41 Lamellar sub-structure developing in titanium irradiated by an electron beam and fractured when fatigue testing. A layer at a depth of 10–30 μm

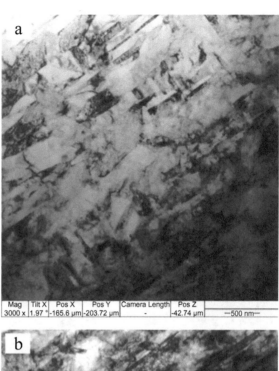

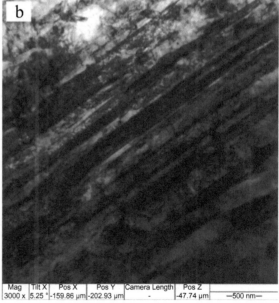

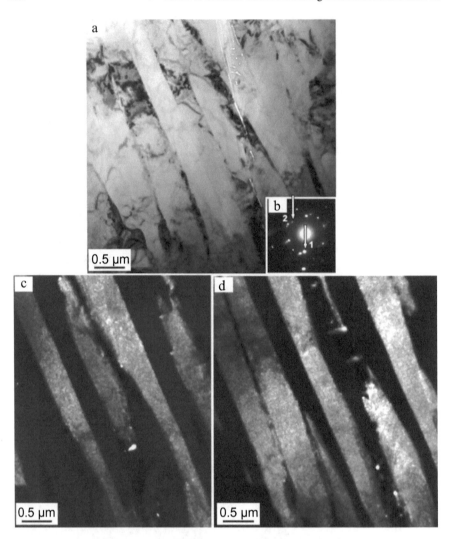

Fig. 9.42 Lamellar structure developed in commercially pure titanium irradiated by an electron beam at a depth of 10–30 μm. (a) bright field; (b) electron-diffraction micro-pattern; (c)(d) dark fields obtained in reflexes [21] α-Ti. The arrows (b) show reflexes, where dark fields were obtained: (c)—reflex 1; (d)—reflex 2

It is suggested that the irradiation of titanium by an intensive electron beam results in the development of a lamellar surface layer (≈ 30 μm). The formation of a lamellar sub-structure facilitated by the high-speed crystallization of a titanium surface layer enhances the fatigue endurance of the material of interest.

The most important finding to emerge from high-cycle fatigue tests of titanium irradiated by a sub-millisecond intensive pulsed electron beam (an energy density of a beam—25 J/cm^2) is the significant growth of its fatigue endurance. It is reported

Fig. 9.43
Electron-microscopic view of a dislocation sub-structure in titanium grains fractured when fatigued. A layer at a distance of ≈ 70 μm below the irradiated surface

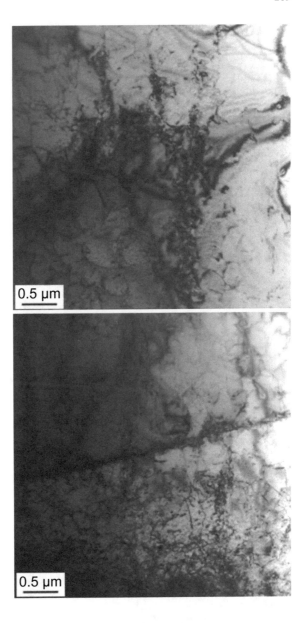

on an approximately 2.2 times increase of the fatigue endurance in the material of interest. Obviously, this effect is attributed to the structural transformations in the titanium surface layer conditioned by the high-speed cooling and heating during electron-beam processing.

A numerical solution to the finding of a temperature field developed in the surface of titanium irradiated by an electron beam provides the data for the assessment of a modified layer thickness, a maximal temperature on the irradiated surface, a temperature gradient, speeds of cooling and heating, as well as time intervals for diverse aggregate states of the material. Given that an energy density of an electron beam is 25 J/cm^2 and an impulse is 150 µs, the irradiation of titanium brings about the melting of the surface with a thickness of 18.1 µm (Fig. 9.44). This layer is divided into two sub-layers: a one-phase (liquid) surface (10.4 µm) (Fig. 9.44, layer 1) and a two-phase (liquid + solid state) intermediate layer (7.7 µm) (Fig. 9.44, layer 2). The first layer exists for 80 µs; a period from the melting of titanium till the completion of crystallization is 147.2 µs.

A morphological study of the titanium fracture surface formed when fatigued after the irradiation by an electron beam reveals a multilayered structure in the volume of a sample close to the irradiated surface (Fig. 9.45).

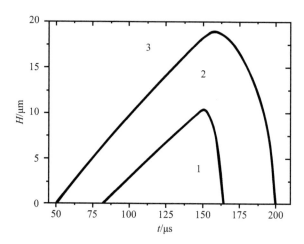

Fig. 9.44 Thickness of layer 1 (liquid) and layer 2 (solid phase + liquid) vs. observation time when processing the titanium surface by an electron beam (energy density 25 J/cm^2, impulse time 150 µs); zone 3—zone of the solid phase location (heat impact layer), reprinted from ref. [2] (Copyright 2016, with permission IOP Publishing Ltd.)

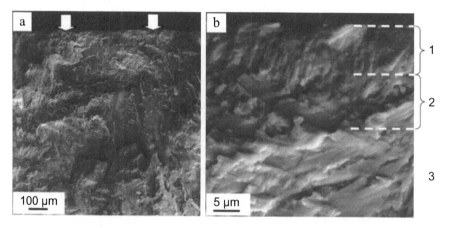

Fig. 9.45 Structure of the fracture surface in the irradiated sample ($E_S = 25$ J/cm^2) when fatigued. The arrows (a) indicate the irradiated surface

9.2 Fracture Surface ...

A columnar structure is detected in a layer of the irradiated surface; being an evidence, therefore, of the high-speed crystallization in the material (Fig. 9.45b). Comparing the data in Figs. 9.44 and 9.45, we assume that surface layer 1 (Fig. 9.45) is a product of melting and high-speed crystallization of a one-phase (liquid) state; layer 2 (Fig. 9.45) developed due to the crystallization of a two-phase (liquid phase + solid phase) state of the material.

The methods of transmission electron microscopy in diffraction mode of thin foils were used in the research into the phase composition and defect sub-structure of surfaces in titanium samples pre-irradiated by an intensive electron beam and fractured during fatigue testing. The research data show the structure of a titanium surface layer irradiated by an intensive electron beam and fractured during fatigue testing differs significantly from the structure of untreated titanium or irradiated at an electron-beam energy density of 25 J/cm^2. For instance, a surface layer (≈ 10 μm) is lamellar (Fig. 9.46a). The lamellae form packages of parallel crystallites and tend to be located at a certain angle to the irradiated surface. From the data of an electron-diffraction microanalysis (Fig. 9.46b) alongside with the dark-fields technique [22], it is seen that lamellae are α-titanium.

At a distance this tendency is recorded less frequently and at a depth of ≈ 10 μm lamellae are spaced more differently relatively to each other and the irradiated surface (Fig. 9.47).

Besides a lamellar structure, a sub-grain one is detected at a depth of 12–15 μm (Fig. 9.48), which starts prevailing in a layer at a depth of 30–40 μm below the irradiated surface (Fig. 9.49).

A grain structure similar to that of untreated samples fractured in fatigue tests is found at a depth ≈ 80 μm and deeper (Fig. 9.50).

According to the results of thermal calculations and data obtained by means of diffraction electron microscopy of thin foils the layers mentioned above might form in fatigue tests owing to the preliminary processing of commercially pure titanium VT1-0 by an intensive pulsed electron beam. For instance, a surface lamellar layer represents a product of the high-speed titanium crystallization from a one-phase (liquid) state; a layer beneath it with a blended (lamellae and sub-grains) structure develops in the crystallization of a two-phase (liquid + solid) state. A layer found deeper is suggested to be in the zone of thermal impact.

Therefore, the study pointed out that the irradiation of titanium by an intensive electron beam and further fatigue tests till the fracture led to the development of a multilayered structure. A layer (≈ 10 μm) close to the irradiated surface displays a lamellar structure only; a layer beneath it (≈ 10 μm) has a blended structure comprising lamellae and sub-grains. A sub-grain structure is a key element in a layer at a depth of 30–40 μm below the irradiated surface (a heat impact layer).

In the study, numerical calculations of a temperature field arising in the titanium surface irradiated by an intensive electron beam were carried out. The important findings show that a lamellar layer close to the irradiated surface is a product of the high-speed crystallization of a one-phase titanium (liquid phase) state, a lamellar—sub-grain sub-layer beneath it develops in the crystallization of a two-phase (liquid + solid phase) state, and a layer with a sub-grain structure is a heat impact layer.

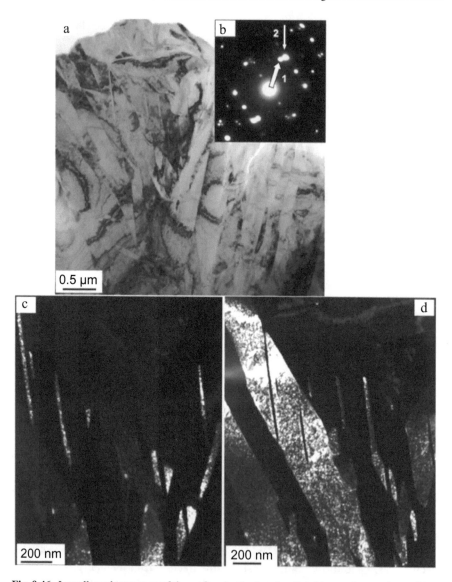

Fig. 9.46 Lamellar sub-structure of the surface in titanium irradiated by an electron beam ($E_S = 25$ J/cm^2) and fractured when fatigue testing. (a) bright field; (b) electron-diffraction micro-pattern; (c)(d) dark fields obtained in reflexes [19] α-Ti (c) and [21] α-Ti (d). The arrows (b) indicate reflexes, where dark fields were obtained: (c)—reflex 1; (d)—reflex 2

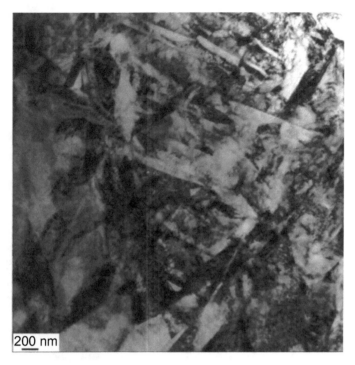

Fig. 9.47 Lamellar sub-structure of the surface—a product of electron-beam processing ($E_S = 25$ J/cm^2) and fatigue fracture. A layer at a depth of ≈ 10 μm

To sum up, the physical reason for the extended fatigue endurance of titanium irradiated by an intensive electron beam is the development of a lamellar structure initiated by the high-speed crystallization of the surface. To summarize, several important conclusions are drawn.

1. The fatigue fracture surface of titanium has been investigated. The investigation has pointed out the development of a thin surface layer, a fatigue fracture zone, and a rupture area. Progression marks at a distance of 3.0 μm between each other have been detected in the fatigue fracture zone. Ductile fracture pits are a principal structural element in the zone of fatigue crack propagation, while quasi-spalling facets dominate in the fracture surface in the rupture area.
2. The fracture of titanium in fatigue tests is a product of the development of a surface layer with a so-called critical structure, unable to the further evolution, i.e. its resource of material plasticity is limited. The nano-crystalline surface layer (≈ 2 μm) contains oxide phase inclusions on boundaries of nano-dimensional α-titanium grains and in their volume.
3. The study has revealed a gradient structure of titanium fractured when fatigued. There is a thin (≈ 2 μm) sub-micro and nano-crystalline multiphase surface layer in it, separated from the main volume by micro-pores and micro-cracks.

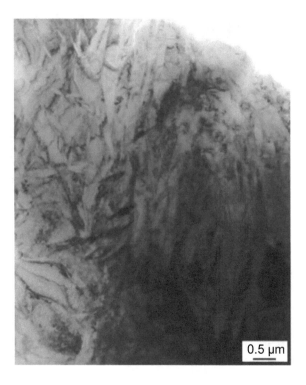

Fig. 9.48 Complex sub-structure made up of lamellae and sub-grains of α-titanium formed as a result of electron-beam processing ($E_S = 25$ J/cm^2). A layer at a depth of 12–15 μm

A layer adjacent to the nano-structured surface layer is determined to be α-titanium with a great number of micro-twins. A thickness of the lamellar layer is as much as 20–25 μm. The number of micro-twins decreases at a distance from the nano-structured layer. Chaotically located dislocations are a principal structural element of the defect sub-structure in the volume of α-titanium grains in a deeper part of a sample.

4. SEM and TEM studies have disclosed a multilayered state in the surface of irradiated samples. The structure of this state depends on a depth below the irradiated surface.
5. The SEM-based research into the fracture surface of titanium pre-irradiated by an intensive sub-millisecond pulsed electron beam (16 keV, 30 J/cm^2, 150 μs, 3 impulses, 0.3 s^{-1}) was carried out. The most important finding is that the irradiation facilitates the formation of a multilayered structure. Furthermore, there are micro-pores parallel to the sample surface in the surface developed when processed by an electron beam.
6. A temperature field arising in the surface of titanium irradiated by an intensive electron beam was numerically calculated. A lamellar layer below the irradiated surface is a product of the high-speed crystallization of a one-phase (liquid phase) titanium state when irradiated by an electron beam; a lamellar and sub-grain sub-layer beneath it develops due to the crystallization of a two-phase (liquid + solid phase) state; and a sub-grain layer is a heat impact layer. Apparently, the physical

9.2 Fracture Surface …

Fig. 9.49 Sub-grain structure of α-titanium formed as a results of electron-beam processing ($E_S = 25$ J/cm^2) and fracturing when fatigued. A layer at a depth of 30–40 μm

reason for the extended fatigue endurance of titanium irradiated by an intensive electron beam is the development of a lamellar sub-structure initiated by the high-speed crystallization of the titanium surface and the reduction of a scalar density of dislocations.

7. Another significant finding is that the irradiation of commercially pure titanium VT1-0 by a high-intensity pulsed electron beam (16 keV, 25 J/cm^2, 150 μs, 3 impulses, 0.3 s^{-1}) leads to the refinement of a grain structure and the formation of an inner-grain sub-structure, i.e. additional sub-micro and nano-dimensional structural elements appear in the surface layer.

8. Finally, the irradiation of titanium in the mode above furthers the formation of a multilayered structure. The study has uncovered the lamellar nature of the surface layer and suggested it to be one of the principal reasons for the significant extending of the fatigue endurance in the material of interest.

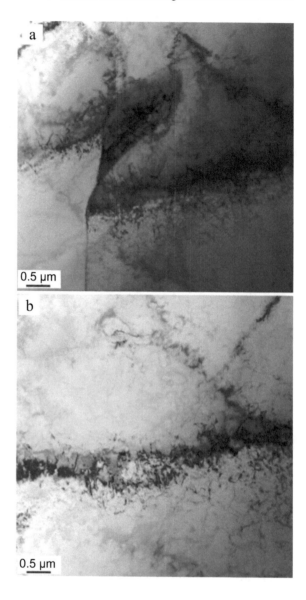

Fig. 9.50 Grain structure in titanium irradiated by an electron beam ($E_S = 25$ J/cm^2) and fractured when fatigue testing. A layer at a depth of ≈ 80 μm

References

1. Konovalov, S.V., et al.: Increase of Fatigue Life of Titanium VT1-0 after Electron Beam Treatment. Key Eng. Mater. **704**, 15–19 (2016)
2. Konovalov, S.V., et al.: Effect of electron beam treatment on structural change in titanium alloy VT-0 at high-cycle fatigue. IOP Conf. Ser. Mater. Sci. Eng. **150**, 012037 (2016)
3. Konovalov, S.V., Kosinov, D.A., Komissarova, I.A., Gromov, V.E.: Effect of electropulsing on the fatigue behavior and change in the austenite steel structure. Materials Science Forum, vol. 906 (2017)

4. Konovalov, S.V., et al.: Structure of titanium alloy, modified by electron beams and destroyed during fatigue. Lett. Mater. **7**, 266–271 (2017)
5. Konovalov, S. et al.: Gradient structure formed in commercially pure titanium irradiated with a pulsed electron beam. In AIP Conference Proceedings, vol. 1909, 020095 (2017)
6. Konovalov, S., et al.: The structure of the surface layer in titanium VT1-0 after high-cycle fatigue tests. IOP Conf. Ser. Mater. Sci. Eng. **447**, 012075 (2018)
7. Konovalov, S., Komissarova, I., Ivanov, Y., Gromov, V., Kosinov, D.: Structural and phase changes under electropulse treatment of fatigue-loaded titanium alloy VT1-0. J. Mater. Res. Technol. **8**, 1300–1307 (2019)
8. Gao, B., et al.: Effect of high current pulsed electron beam treatment on surface microstructure and wear and corrosion resistance of an AZ91HP magnesium alloy. Surf. Coatings Technol. **201**, 6297–6303 (2007)
9. Ivanov, Y.F., et al.: Formation of nanocomposite layers at the surface of VT1-0 titanium in electroexplosive carburization and electron-beam treatment. Steel Transl. **42**, 499–501 (2012)
10. Sizov, V.V., Gromov, V.E., Ivanov, Y.F., Vorob'ev, S.V., Konovalov, S.V.: Fatigue failure of stainless steel after electron-beam treatment. Steel Transl. **42**, 486–488 (2012)
11. Gromov, V.E., Gorbunov, S.V., Ivanov, Y.F., Vorobiev, S.V., Konovalov, S.V.: Formation of surface gradient structural-phase states under electron-beam treatment of stainless steel. J. Surf. Investig. X-ray, Synchrotron Neutron Tech. **5**, 974–978 (2011)
12. Ivanov, Y.F., et al.: Multicyclic fatigue of stainless steel treated by a high-intensity electron beam: surface layer structure. Russ. Phys. J. **54**, 575–583 (2011)
13. Sizov, I., Mishigdorzhiyn, U., Leyens, C., Vetter, B., Fuhrmann, T.: Influence of thermocycle boroaluminising on strength of steel C30. Surf. Eng. **30**, 129–133 (2014)
14. Panin, V.E., Panin, A.V.: Effect of the surface layer in a solid under deformation. Fiz. Mezomekhanika **8**, 7–15 (2005)
15. Bagmutov, V.P., Parshev, S.N.: Integrated approach to the electromechanical formation of a structurally inhomogeneous surface layer on steel parts. Steel Transl. **34**, 66–69 (2004)
16. Bagmutov, V.P., Parshev, S.N., Polozenko, N.Y.: Improvement of mechanical characteristics of a cutting edge of the blade instrument by means of electromechanical processing. Mechanika **56**, 18–20 (2005)
17. Bagmutov, V.P., Parshev, S.N.: Integrated concept of formation of structurally nonuniform surface layer of steel products by electromechanical treatment. Izv. Ferr. Metall. 69–71 (2004)
18. Maruschak, P.O., Mocharskyi, V.S., Zakiev, I.M., Nikiforov, Y.M.: Morphology of periodical structures on surface of steel 15Kh13MF after the nanosecond laser irradiation accompanied by generation of shock waves. In 2012 IEEE International Conference on Oxide Materials for Electronic Engineering (OMEE), pp. 192–193 (IEEE, 2012). https://doi.org/10.1109/omee.2012.6464902
19. Zhang, K.M. et al.: Surface modification of Ni (50.6at.%) Ti by high current pulsed electron beam treatment. J. Alloys Compd. 434–435, 682–685 (2007)
20. Uglov, V.V., et al.: Structure, phase composition and mechanical properties of hard alloy treated by intense pulsed electron beams. Surf. Coatings Technol. **206**, 2972–2976 (2012)
21. Ivanova, V.S., Goritskii, V.M., Orlov, L.G., Terent'ev, V.F.: Dislocational structure of iron at the tip of a fatigue crack. Strength Mater. **7**, 1312–1317 (1975)
22. Fenner, D.B., Hirvonen, J.K., Demaree, J.D.: Selected topics in ion beam surface engineering. Engineering Thin Films and Nanostructures with Ion Beams. (2005)